2020 国际大学生建筑和结构设计竞赛作品集

主　　编　丁　帅
副 主 编　黄　鹤　黄　兴
参　　编　李　洋　刘　星　刘书青
　　　　　王　璇　田紫嫣　刘　璐
组织编写　“一带一路”建筑类大学国际联盟秘书处

中国建材工业出版社

图书在版编目（CIP）数据

2020国际大学生建筑和结构设计竞赛作品集 / 丁帅主编. -- 北京 : 中国建材工业出版社, 2021.11
ISBN 978-7-5160-3322-7

Ⅰ. ①2… Ⅱ. ①丁… Ⅲ. ①建筑结构－结构设计－作品集－中国－现代 Ⅳ. ①TU318

中国版本图书馆CIP数据核字（2021）第199905号

内容简介

"一带一路"建筑类大学国际联盟秘书处编辑的《2020国际大学生建筑和结构设计竞赛作品集》一书，共收录竞赛征集的来自6个国家17所高校的50件作品。其中建筑设计方向作品27件，土木结构方向作品23件。

竞赛以"一带一路·建设未来"为主题，分为建筑设计、土木结构两个方向，分别聚焦"历史城区中的未来图书馆设计"及"融入周边环境的特色桥梁设计"。组委会聘请中国工程院院士、全国工程勘察设计大师崔愷院士为特聘专家进行指导，邀请国内外建筑、土木学科知名专家组成了竞赛作品评审专家委员会。竞赛为"一带一路"建筑类大学搭建了科技创新与教育交流平台，有效助力国际化创新型专业人才的培养，同时也为参赛学生提供了良好的实践机会和沟通融合的平台，引导、促进了相关专业的教育教学改革。

本书可供广大工科院校师生以及建筑设计师、结构工程师、城市设计研究人员参考。

2020国际大学生建筑和结构设计竞赛作品集
2020 Guoji Daxuesheng Jianzhu he Jiegou Sheji Jingsai Zuopinji
丁　帅　主编

出版发行：中国建材工业出版社
地　　址：北京市海淀区三里河路1号
邮　　编：100044
经　　销：全国各地新华书店
印　　刷：北京天恒嘉业印刷有限公司
开　　本：889mm×1194mm　1/16
印　　张：11
字　　数：200千字
版　　次：2021年11月第1版
印　　次：2021年11月第1次
定　　价：198.00元

本社网址：www.jccbs.com，微信公众号：zgjcgycbs
请选用正版图书，采购、销售盗版图书属违法行为

前　言

Foreword

为积极响应国家“一带一路”倡议，深入贯彻落实北京市《新时期北京教育对外开放工作规划（2016—2020）》《北京“十三五”时期加强国际交往中心建设规划》和《北京市对接共建“一带一路”教育行动计划实施方案》等文件精神，北京建筑大学发起并于 2017 年 10 月 10 日成立了“一带一路”建筑类大学国际联盟（以下简称“联盟”）。

目前，俄罗斯、波兰、法国、美国、英国、亚美尼亚、保加利亚、捷克、韩国、马来西亚、希腊、尼泊尔、以色列和中国等 27 个国家的 64 所大学成为联盟成员高校。北京建筑大学作为联盟主席单位，通过积极谋划和顶层设计，不断创新推动联盟的可持续发展。

2020 年，联盟首次组办了“2020 国际大学生建筑和结构设计竞赛”。竞赛以“一带一路·建设未来”为主题，其宗旨是培养大学生的科技创新精神和实践能力，提高大学生科学素养和科研技能水平，为增强各国大学生之间的学术和文化交流，促进“一带一路”建筑类大学的交流合作搭建平台。来自国内外 18 所高校、50 支参赛队伍、250 余名师生针对“历史城区中的未来图书馆”和“融入周边环境的特色桥梁设计”的题目开展深入调研和创新设计，充分展现了“一带一路”沿线不同国家的文化内涵、历史风貌和风土人情。

全书根据两大题目，分为“建筑设计方向”和“土木结构方向”两章，每个设计作品采用图文并茂的方式呈现。[①]

在建筑设计方向，围绕“历史城区中的未来图书馆”这一主题，鼓励学生在老城中设计一座“未来图书馆”，将图书馆及其环境看作与城市景观、社会、文化、公共生活、可持续性等要素相融合的“生态系统”，期待能将图书馆作为催化剂，为城市更新提供新的视野。本章所收录的 27 件作品，是学生们创意设计展现出的坐落在“一带一路”沿线国家古老历史街区里

① 书中部分图片做虚化处理，如有疑问，欢迎联系“一带一路”建筑类大学国际联盟秘书处。

的风格迥异、充满创意的图书馆：北京建筑大学的师生们选取北京西海胡同、前门大街历史街区，打造出散落在胡同内的图书角；重庆交通大学的师生们设计的“浮屋阅栈”，坐落于重庆江津区白沙古镇的朝天嘴码头；波兰华沙生态与管理大学的同学们畅想在华沙普拉斯基港畔建设一座代表城市与自然和谐的绿色图书馆；亚美尼亚国立建筑大学的师生们走进亚美尼亚圣拉扎罗岛，为当地独具特色的图书馆提供各式各样的改建方案；俄罗斯莫斯科理工大学梁赞分校设计的图书馆位于梁赞旧城区的特鲁别日河岸边，靠近克里姆林宫；尼泊尔加德满都大学则在尼泊尔加德满都郊区的基尔蒂普尔古镇设计出一片“地下空间”，突出将未来空间与过去相连接的概念……

在土木结构方向，以“融入周边环境的特色桥梁设计”为主题，建议学生们选择合适的跨越道路路线、江河湖海、深山峡谷等地貌的桥址，在遵循安全、适用、耐久、创新、美观、环保、节能等设计理念的前提下设计公路桥、铁路桥、市政桥、景观桥、人行桥、自行车桥、综合桥等。本章收录了 23 件土木结构方向作品：长安大学的同学们进行了“跨国设计”，为“一带一路”沿线国家巴西设计了一座气势磅礴的跨海大桥；北京建筑大学的师生团队选择了地处雄安新区启动区的一处新建桥梁作为设计对象；韩国江原大学的同学们设计了连通济州岛和牛岛的标志性桥梁；莫斯科国立建筑大学的同学们为移民火星做了超前准备，设计了外星移民用桥梁，万一有类似情况发生时可以帮助探测器及其他装置通过峡谷和环形山；山东建筑大学的同学们设计了双塔双索面斜拉桥“弦歌桥”；亚美尼亚国立建筑大学的同学们为旅游热门城市杰克穆勒设计了一座多功能桥梁酒店……

竞赛专家委员会由 17 位中外专家组成，其中中国工程院院士崔愷教授为特聘专家，北京建筑大学建筑与城市规划学院院长张杰教授为组长，东伦敦大学建筑、计算机及工程学院院长谭大维教授为副组长，建筑设计与土木结构两个方向各 7 名专家为评审委员。

“建筑教育需要激发学生的创造力、艺术潜能和空间想象力，国际交流和交往在其中发挥着重要的作用，而举办国际竞赛是促进各国师生沟通交流的一个途径。通过竞赛，大家相互传授知识和相互学习借鉴，填补各校在学术研究方面的空白，促使大家对课程进行定期更新以跟上世界最前沿的趋势。”亚美尼亚国立建筑大学代理校长叶吉扎尔·瓦尔达尼扬（Yeghiazar Vardanyan）教授在贺信里说。

本次竞赛无疑会成为国际大学生之间的学术和文化交流盛会，中外大学生们置身于“一带一路”沿线国家建设的前沿，用自己的专业知识和创意思维，践行民心互通、政治互信、文化包容的“一带一路”倡议。在此，对联盟成员高校的积极参与及广泛支持表示衷心感谢！

编　者

2021.10.1

Foreword

To actively respond to China's Belt and Road Initiative and *Beijing's Work Plan for Opening Up Education in the New Period (2016—2020), Beijing's Thirteenth Five Year Plan for Improving Its Role as an International Exchange Center, Beijing's Implementation Plan for Participating in Promoting Education Along the Belt and Road*, the Belt and Road Architectural University International Consortium (BRAUIC) was initiated by the Beijing University of Civil Engineering and Architecture (BUCEA) on October 10, 2017.

BRAUIC comprises 64 universities hailing from 27 countries, including Russia, Poland, France, the United States, the United Kingdom, Armenia, Bulgaria, the Czech Republic, Korea, Malaysia, Greece, Nepal, Israel and China. As the chairman's work unit of the BRAUIC, BUCEA is committed to promoting the sustainable development of the BRAUIC through active planning and effective structural design.

In this context, the BRAUIC organized the First International Student Architectural and Structural Design 2020 under the theme of "Build the Belt and Road, Build the Future". The competition was designed to cultivate the scientific and technological innovation spirits and practical abilities of university students and enhance their scientific literacy and research skills, serving as a platform for the academic and cultural exchange among students from different countries, and the exchanges and cooperation among BRAUIC members. Altogether, over 250 participants across 50 teams hailing from 18 universities from all over the world submitted their entries under two categories: "Future Library in A Historical Urban Area" and "Characteristic Bridge Design Fitting into the Surroundings". The submissions, born out of participants' in-depth research and innovative design, fully demonstrated the cultural, historical, and humanistic elements of countries along the Belt and Road.

Aaccording to the two categories above, there will be two chapters in the book, each design-work is presented with introductions and pictures.[①]

For the architectural design category themed on "Future Library in a Historical Urban Area", participants were challenged to design "a future library" for an ancient city. They were encouraged to take the library and its environment as an "ecosystem" integrated with the urban landscape, society, culture, public life, sustainability, and other elements. In so doing, the library would serve as a catalyst that expanded the horizon for urban renewal. The 27 submissions in this chapter were a testament to the creativity of their designers. The libraries varied in styles and originality while blending in with historic streets in countries along the Belt and Road. Among these designs, participants from BUCEA, China, turned to the Xihai Hutong and the historic

① Some feather pictures are permitted in this book, any question, please contact BRAUIC.

Qianmen district in Beijing, creating scattered reading corners nested within traditional hutongs; participants from Chongqing Jiaotong University, China, designed a "Floating Library" for the Chaotianzui Wharf in the historic Baisha Ancient Town of Jiangjin District in Chongqing; participants from the University of Ecology and Management in Warsaw, Poland, envisioned a green library on the bank of Port Praski to create a balanced ecosystem which fits in the urban landscape; participants from the National University of Architecture and Construction of Armenia set their sights on Saint Lazarus of the Armenians and provided a series of renovation plans for the unique local library; the library designed by participonts from the Ryazan Institute (branch) of Moscow Polytechnic University in Russia was located in the historical part of Ryazan near the Kremlin on the bank of the Trubezh River; while "The Chamber Under" submitted by participonts from Kathmandu University of Nepal designed for Kirtipur, an ancient town on the outskirts of Kathmandu, highlighted the connection between a future space and the past……

In the structural design category, the students were tasked with designing a "Characteristic Bridge Design Fitting into the Surroundings". Students were advised to select appropriate bridge sites that spanned road routes, rivers, lakes, seas, mountains, and valleys. The bridge types included but were not limited to highway bridges, railway bridges, municipal bridges, landscape bridges, pedestrian bridges, bicycle bridges, and comprehensive bridges, presenting such design concepts as safety, applicability, durability, innovation, beauty, environmental protection, energy conservation, etc. This chapter contained 23 submissions: the participonts of Chang'an University in China envisioned "a transnational construction" in which they designed a magnificent cross-sea bridge connecting Salvador and Itaparica for Brazil, a country along the Belt and Road; the BUCEA team selected a new bridge construction site in the Xiong'an New Area's start-up zone to base their designs; students from Kangwon National University, Korea, designed a TEU-bridge to connect Jeju Island with Udo Island; while students of Moscow State University of Civil Engineering, Russia, designed a bridge for immigrating to Mars, making advanced preparations for the immigration by finding ways to help detectors and other equipment traverse canyons and craters; students of Shandong Jianzhu University in China submitted a double-tower double-cable-plane cable-stayed bridge named the "Xiange Bridge"; students of the National University of Architecture and Construction of Armenia designed a multi-functional bridge with hotel and multifunctional complex in Jermuk, a popular resort city……

The submissions were judged by the Scientific Committee for the Competition, an expert panel of 17 experts from both home and abroad, including Honorary Member Professor Cui Kai, Academician of the Chinese Academy of Engineering; Professor Zhang Jie, Dean of the School of Architecture and Urban Planning of BUCEA, serving as Chair of the panel; Professor David Tann, Dean of the School of Architecture Computing and Engineering of the University of East London, the UK, serving as Deputy Chair. Each of the two categories, the architectural design and structural design, had 7 experts as a jury.

"It is important to activate the student's spatial imagination. In this regard, in addition to formal classes, the educational contacts on the international level are also very important. The best manifestations of this are the contests, especially the international ones, where teaching and learning from each other become the best way of becoming a professional. This, apparently, fills the gap we have in our academic programmes and for the modernization of which we often update them." stated Professor Yeghiazar Vardanyan, Rector of National University of Architecture and Construction of Armenia, in his congratulatory letter.

The competition has undoubtedly become a pageant of academic and cultural exchanges for university students all over the world. With their professional knowledge and creative mindset, these students demonstrated the spirit of the Belt and Road Initiative: people-to-people bonds, political mutual trust, and cultural inclusiveness at the development forefront of the Belt and Road countries. Taking this opportunity, we would like to extend our sincere gratitude to the members of the BRAUIC for their active participation and extensive support!

Compiler

October 1, 2021

目录

Conten

建筑设计方向
Category A: Architectural Design

■ 优秀奖 Honorable Mention

土木结构方向
Category B: Structural Design

■ 优秀奖 Honorable Mention

■ 其他作品 Other Works

建筑设计方向

Category A: Architectural Design

建筑设计方向——历史城区中的未来图书馆

随着互联网的广泛应用，传统图书馆的功能已逐渐动摇。在一个数字信息无处不在的世界里，未来的图书馆可能扮演什么样的角色？在城市空间中的定位和作用又将如何？

与此同时，历史城区的更新是世界范围内被广泛探讨的议题。这些城区建成年代久远，已无法满足居民日益发展的生活需求，除了程度不同的物质性老化外，其功能性、结构性和与城市的关系也暴露出越来越多的问题。因此我们期待，将图书馆作为催化剂，为城市更新提供新的视野。

鉴于此，我们鼓励学生在更综合的背景下去思考老城中的未来图书馆，将图书馆及其环境看作与城市景观、社会、文化、公共生活、可持续性等要素相融合的“生态系统”，并提出相应的、富有想象力的设计方案。

Category A (Architectural Design): Future Library in A Historical Urban Area

With the wide application of the Internet, the function of traditional libraries has become weakened gradually. In a digital and information-based world, what could be the role of a future library? And what will be its positioning and function in urban space?

At the same time, the renewal of historical urban area is a topic widely discussed around the world. These urban areas have been built for a long time and can no longer meet the growing needs of residents. In addition to different degrees of physical aging, more and more problems have emerged in its functionality, structure and relationship with the city. Therefore, we look forward to using the library as a catalyst to provide a new vision for urban renewal.

In view of this, we encourage students to put a future library in an ancient city under a more comprehensive context, take the library and its environment as an “ecosystem” integrated with urban landscape, society, culture, public life, sustainability and other elements, and propose a creative design scheme accordingly.

■ 一等奖 Frist Prize

浮屋阅栈

Floating Library

学校名称 University (College) Name
重庆交通大学，中国
Chongqing Jiaotong University, China

指导教师 Supervisor(s)
温 泉 Wen Quan　　许 可 Xu Ke

参赛学生 Participant(s)
唐 佳 Tang Jia　　李雪晴 Li Xueqing
金圣悦 Jin Shengyue　　祁建华 Qi Jianhua
唐 浩 Tang Hao

设计说明 Design Description

该设计从人群行为本身出发，不拘泥于传统的内部空间布局形式，探索未来图书馆可能的体验式阅读，设置了五根轴线通道作为不同主题的展陈空间，以白沙镇的历史线为中心，左右两侧设置另外四种体验式阅读主题，创造不同情境的沉浸式阅读体验。从建筑形态上来说，在结合材料结构属性与功能属性的同时，转译传统民居建筑的特点。色彩、对山地地形的回应、连绵的屋面、灰空间等都体现于此。

该图书馆不仅仅具有室内的使用功能，也在端部设置了亲水栈道。更新码头文化的同时，也更希望其具有公共开放性。打破固有的内向性图书馆印象，使空间能够流动，视线能够沟通，内外能够交融，使得该建筑与外部空间环境进行交互对话，激发人群更多样性的行为，提升空间活力，营造古镇公共交往的文化氛围。

效果图

场地现状分析 Site Status Analysis

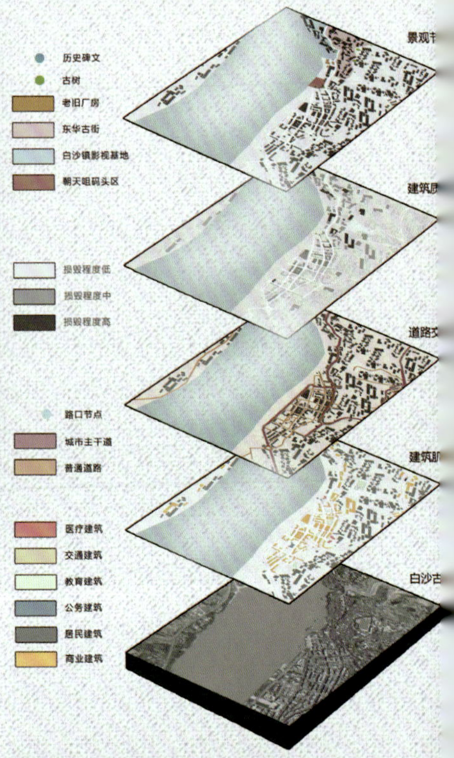

历史进程 Historical Process

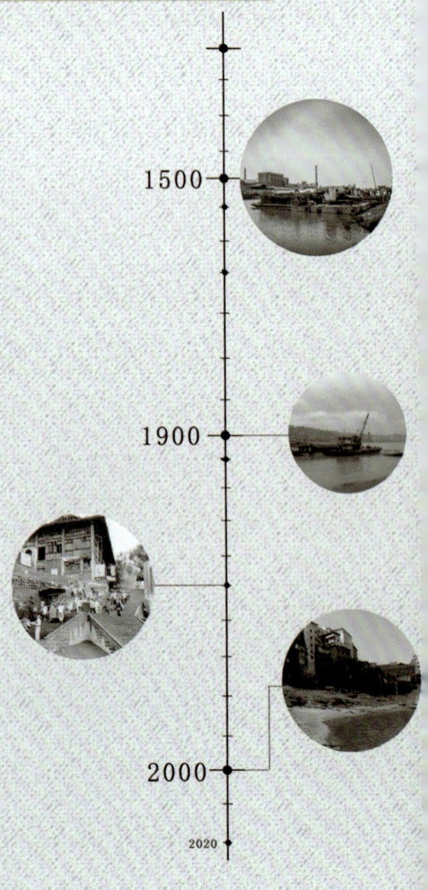

Starting from the crowd behavior itself, the design does not stick to the traditional internal space layout form, instead, it explores the possibility of experiential reading in the future. There are five axis channels as exhibition space of different themes. Taking the historical line of Baisha Town as the center, the library sets four other experiential reading themes on the left and right sides to create immersion reading experience in different situations. In terms of architectural form, this library not only combines the structural and functional properties of materials, but also the characteristics of traditional residential buildings is translated. The color, the response to mountainous terrain, the continuous roofs and gray space are all reflected in this.

The library not only has indoor functions, but also sets up a hydrophilic plank road at the end. While updating the dock culture, we also hope that it will be public and open. Our design plan Breaks the inherent impression of introverted library, which the space can flow, the line of sight can be communicated, and the internal and external can be blended. This solution can let the building having an interactive dialogue with the external space environment, stimulating more diverse behaviors of the crowd, enhancing the vitality of the space, and creating a cultural atmosphere of public communication in the ancient town.

区位分析 Location Analysis

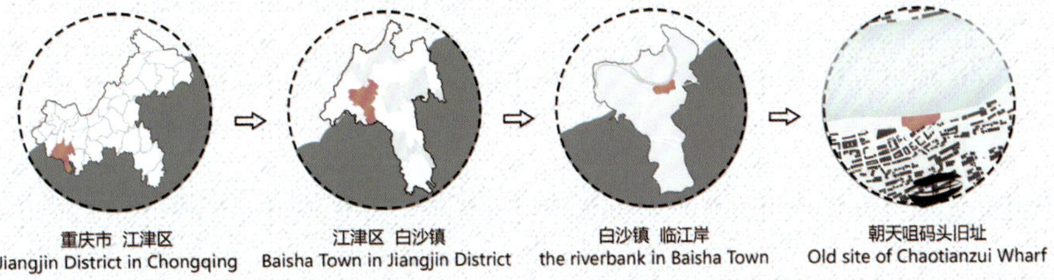

概念生成 Concept Generation

场地实景&元素提取 Site Status & Element Extraction

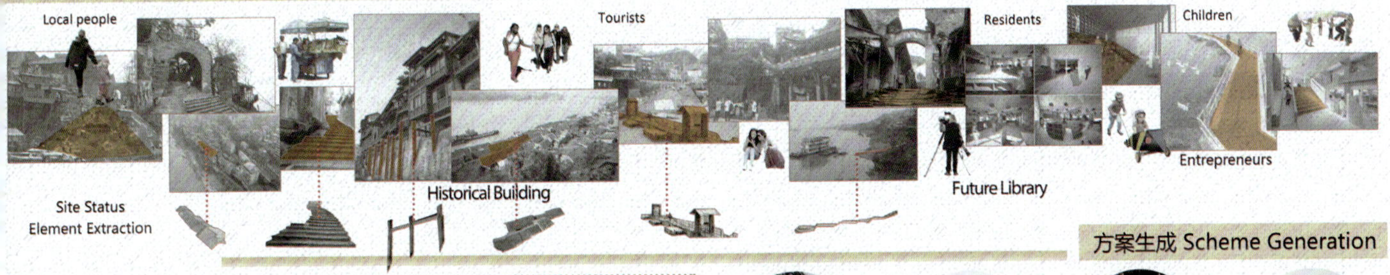

方案生成 Scheme Generation

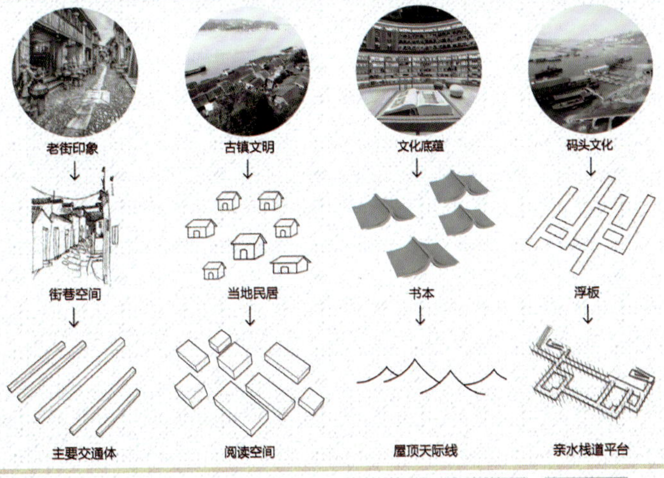

设计说明

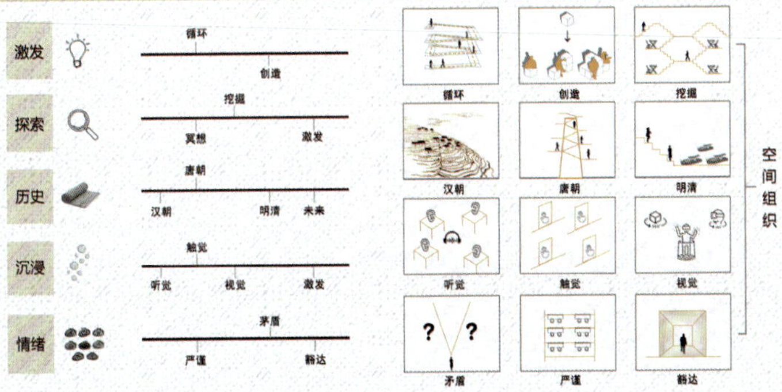

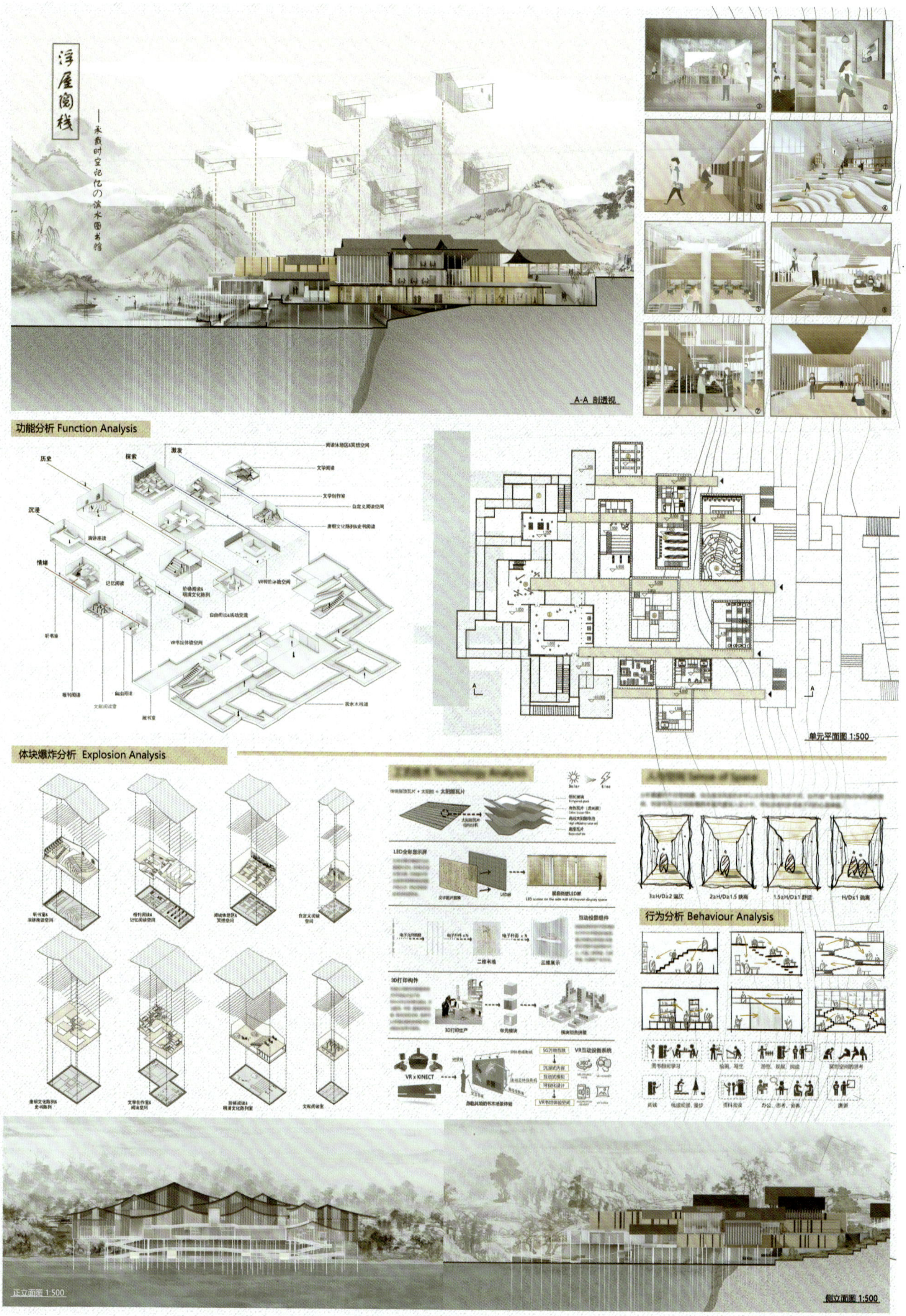
浮屋阁栈
——承载时空记忆の淡水图书馆
A-A 剖透视
功能分析 Function Analysis
历史
探索
激发
沉浸
情绪
单元平面图 1:500
体块爆炸分析 Explosion Analysis
行为分析 Behaviour Analysis
LED全彩显示屏
3D打印构件
VR x KINECT
正立面图 1:500
侧立面图 1:500

■ 二等奖 Second Prize

变·动

Change

学校名称 University (College) Name
安徽建筑大学，中国
Anhui Jianzhu University, China

指导教师 Supervisor(s)
徐丽萍 Xu Liping

参赛学生 Participant(s)
程昕然 Cheng Xinran
贾慧琳 Jia Huilin
阮思琦 Ruan Siqi
王之璐 Wang Zhilu
徐 图 Xu Tu

设计说明 Design Description

随着时代的变化发展，传统的图书馆已经满足不了人们日益丰富的精神和文化需求。群众需要的是聚会的场所、体验的世界、自由的空间。除了图书馆本应具有的阅读功能外，使用者还可以体验更多文化相关类活动。

我们的设计，挑战建筑的"底线"，把不变的建筑，让它外形变化多端，且随着季节时间的变化位置是不固定的，内部功能也不是一成不变的。为了便于搬运变动整体造型结构，采用集装箱的形式，可以通过小体块之间的拼接、堆叠而形成不同的建筑形态。

With the change and development of the times, the traditional library has been unable to meet people's increasingly rich spiritual and cultural needs. What the need of maxes is a place for gathering, a world to experience, and a free space. Besides the reading function that the library should have, users can also experience more cultural related activities.

Our design challenges, the "bottom line" of architecture, changing the shape of the building without change, moreover, with the change of seasons, the position is not fixed, and the internal function is not fixed. In order to facilitate the handling and changing, the overall modeling structure is in the form of containers, which can form different architectural forms by splicing and stacking from small blocks.

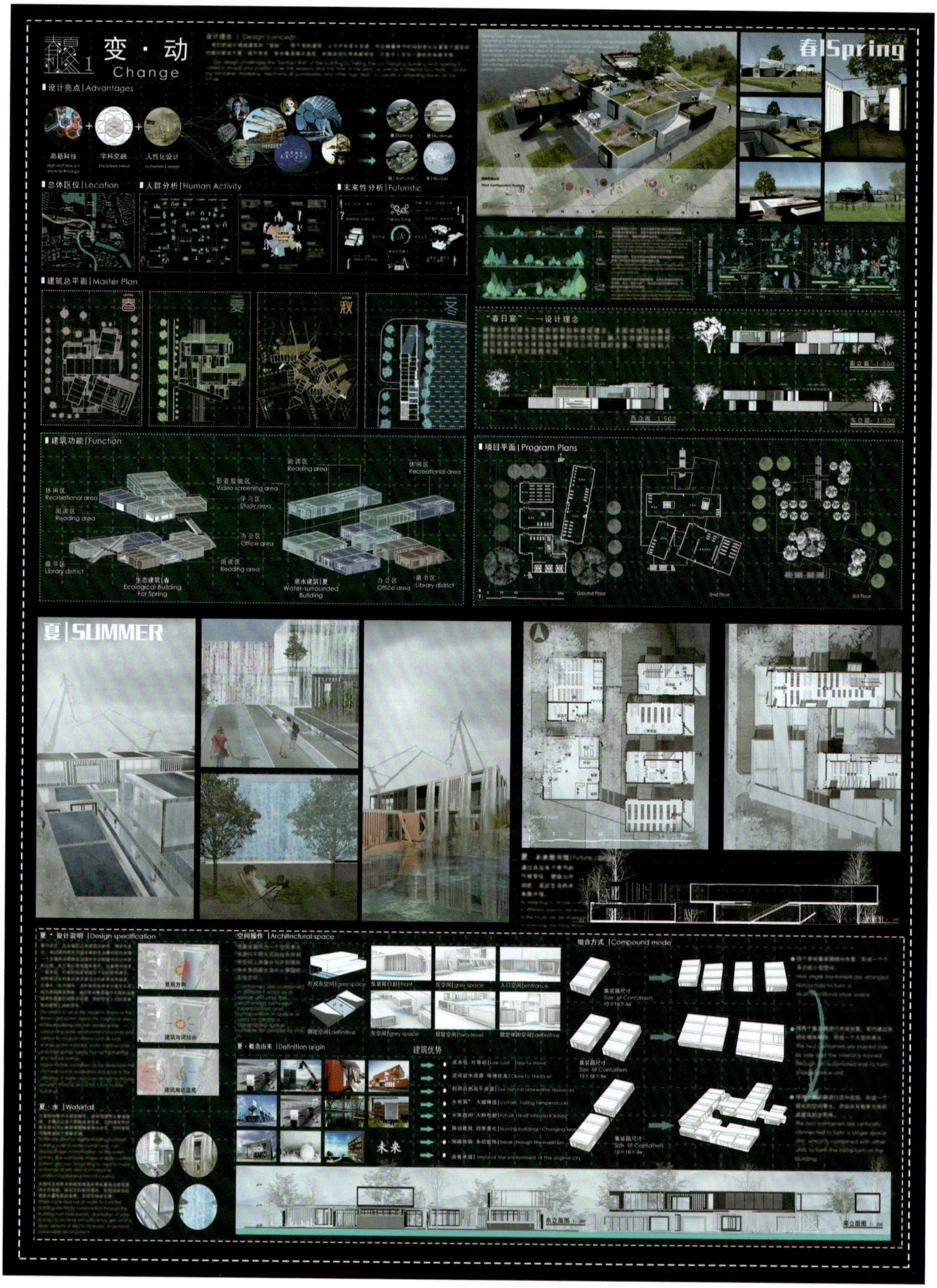
变·动
Change
设计亮点 | Advantages
高新科技
学科交融
人性化设计
总体区位 | Location
人群分析 | Human Activity
未来性分析 | Futuristic
建筑总平面 | Master Plan
春
夏
秋
冬
春 | Spring
"春日宴" —— 设计理念
建筑功能 | Function
阅读区
Reading area
休闲区
Recreational area
影音放映区
Video screening area
学习区
Study area
办公区
Office area
藏书区
Library district
生态建筑 | 春
Ecological Building For Spring
亲水建筑 | 夏
Water-surrounded Building
项目平面 | Program Plans
Ground Floor
2nd Floor
3rd Floor
夏 | SUMMER
夏·设计说明 | Design specification
空间操作 | Architectural space
组合方式 | Compound mode
夏·水 | Waterfall
夏·概念由来 | Definition origin
建筑优势
未来

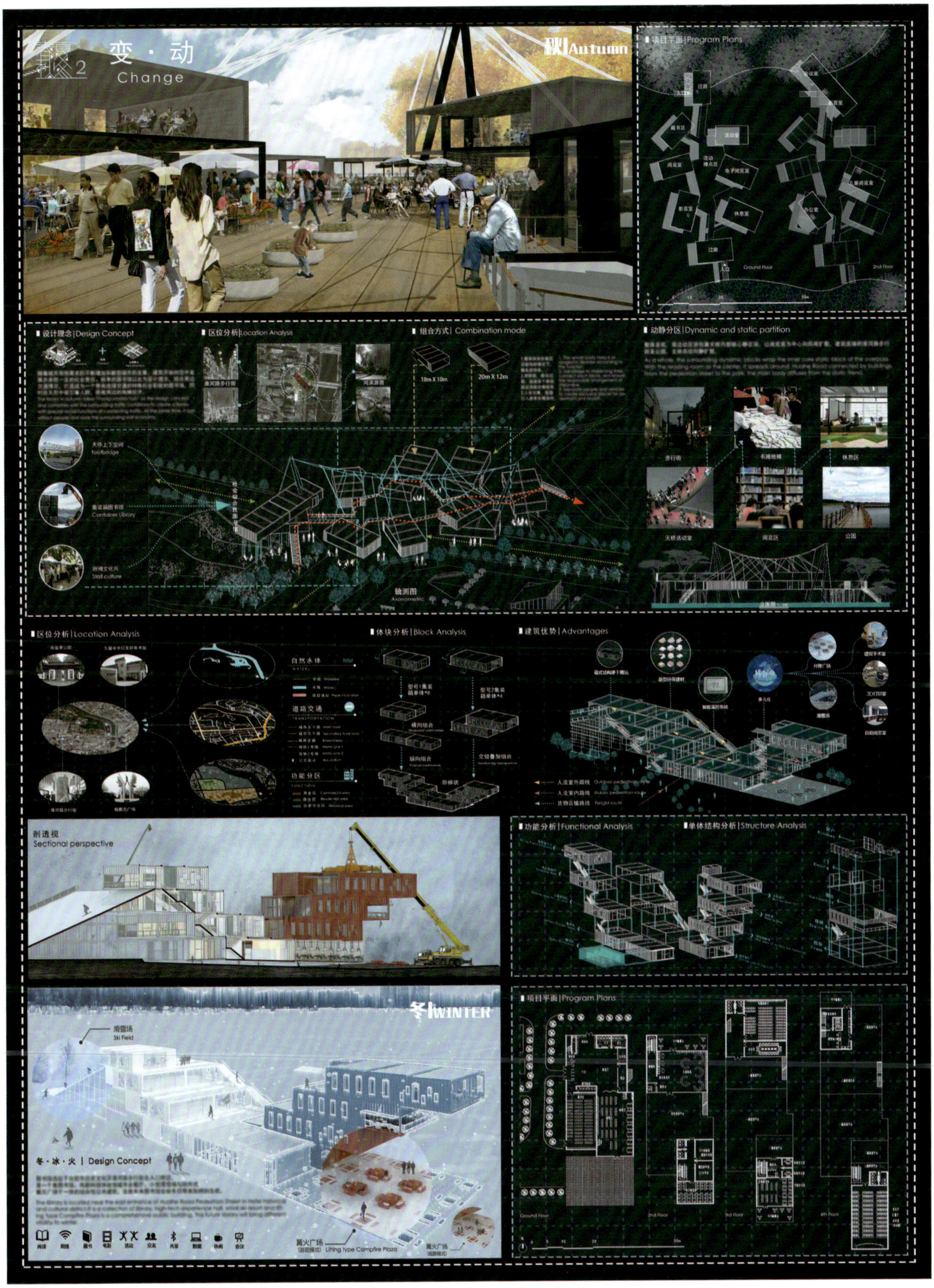
变·动
Change
秋|Autumn
项目平面|Program Plans
设计理念|Design Concept
区位分析|Location Analysis
组合方式|Combination mode
18m X 10m
20m X 12m
动静分区|Dynamic and static partition
天桥上下空间
footbridge
集装箱图书馆
Container library
摊铺文化类
Stall culture
轴测图
Axonometric
步行街
休息区
阅览区
公园
区位分析|Location Analysis
体块分析|Block Analysis
建筑优势|Advantages
自然水体
道路交通
功能分区
剖透视
Sectional perspective
功能分析|Functional Analysis
单体结构分析|Structure Analysis
冬|WINTER
滑雪场
Ski field
冬·冰·火 | Design Concept
篝火广场
Lifting type Campfire Plaza
项目平面|Program Plans
Ground Floor
2nd Floor
3rd Floor
4th Floor

■ 二等奖 Second Prize

西海摘要

Extract of the West Sea

学校名称 University (College) Name
北京建筑大学，中国
Beijing University of Civil Engineering and Architecture, China

指导教师 Supervisor(s)
任中琦 Ren Zhongqi

参赛学生 Participant(s)
黄锐展 Huang Ruizhan　马一丁 Ma Yiding
杨虎臣 Yang Huchen　赵维珩 Zhao Weiheng

设计说明 Design Description

随着社会的发展，北京内城老旧房屋的问题越来越明显。毗邻后海的这一部分房屋，由于这座图书馆的出现，将旧城那些曾经的文化、传统的记忆再次凝聚在一起，将为这片区域重新焕发活力。

未来的图书馆可能不再有很强的藏书功能，不再仅仅是一本本书的集合，更是一个微社会；不仅仅是知识摄取的地方，更是一个人们交往、认识历史与文化、信息交换的一个载体。建筑本身就会更像是一本饱满的书，它也会是一条纽带，让旧城与未来发生交集。

With the development of society, the problems of old houses in Beijing's inner city have become increasingly prominent. Thanks to the emergence of this library, some houses around Houhai are able to gather the cultural and traditional memories of the old city again to rejuvenate this area.

A library in the future may no longer be a place merely dealing books. It will be a micro community for knowledge acquisition to learn about history and culture, and for interaction with people as a carrier of information exchange. The building itself will be more like a book full of contents. It will also be a link connecting the old city with the new world.

西海摘要——散落在胡同内的图书角

Extract of the West Sea

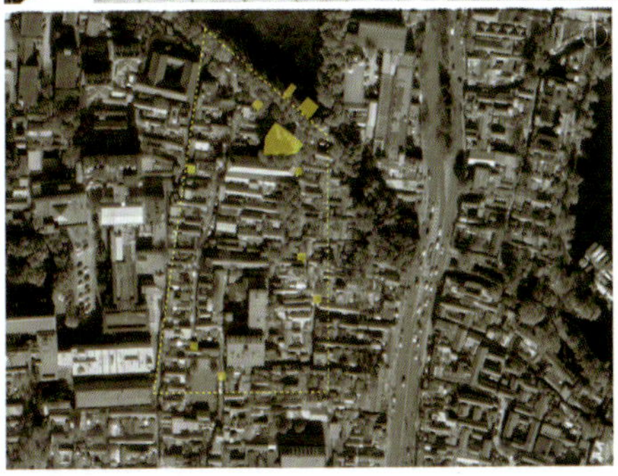

设计说明：

Design Description

❶ 养鸽子 Pigeon Cage

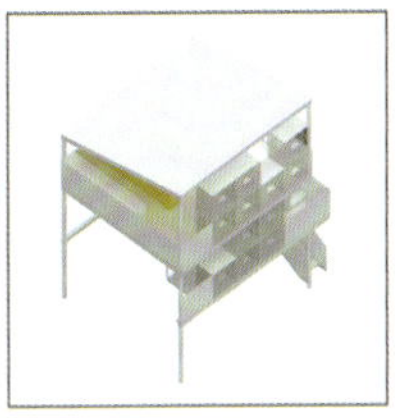

❷ 遛鸟 Bird walking

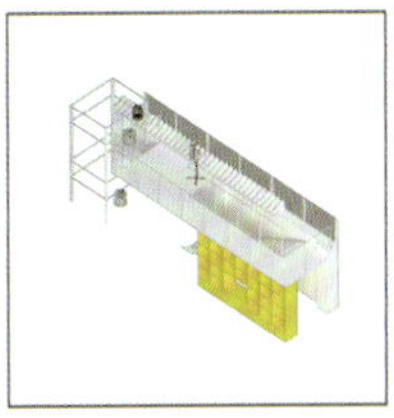

❸ 种菜 Vegetable garden

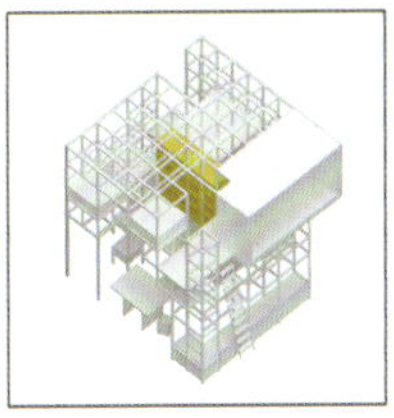

❹ 快递 Express Point

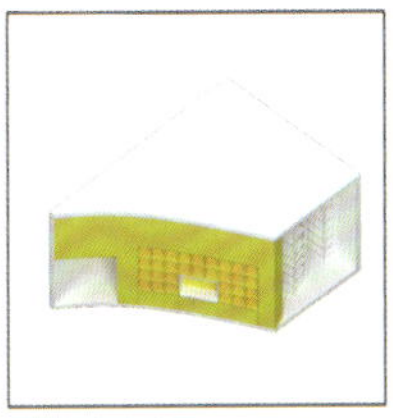

❺ 观景 Observing the scenery

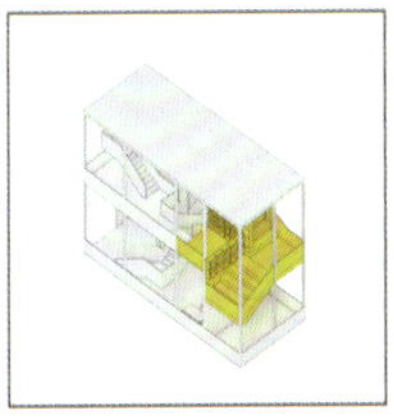

❻ 儿童活动 Kid's Activities

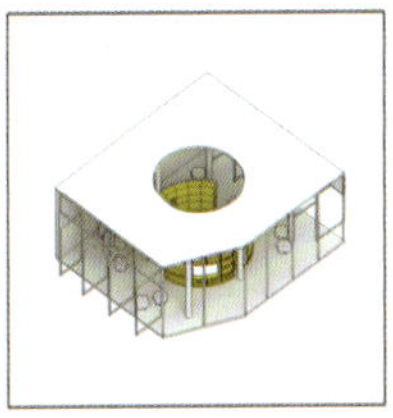

❼ 晒被子 Airing Quilts

❽ 自行车停放 Parking Bicycle

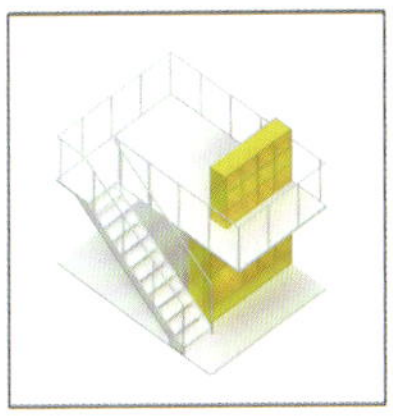

西海摘要——散落在胡同内的图书角

Extract of the West Sea

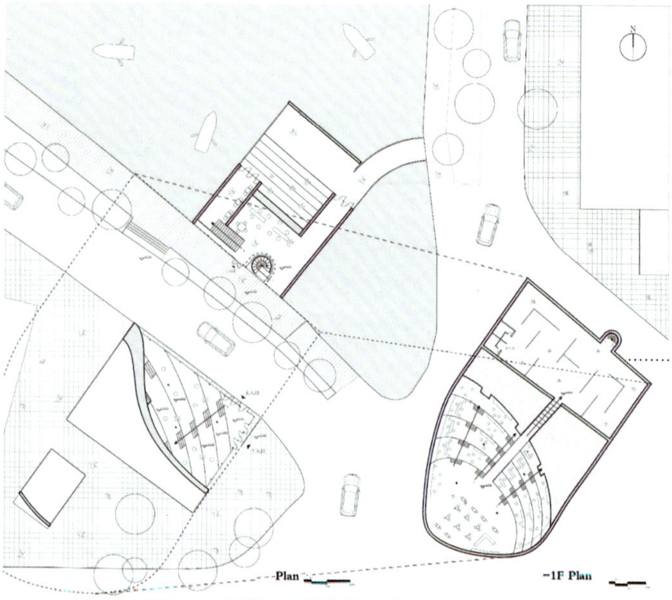

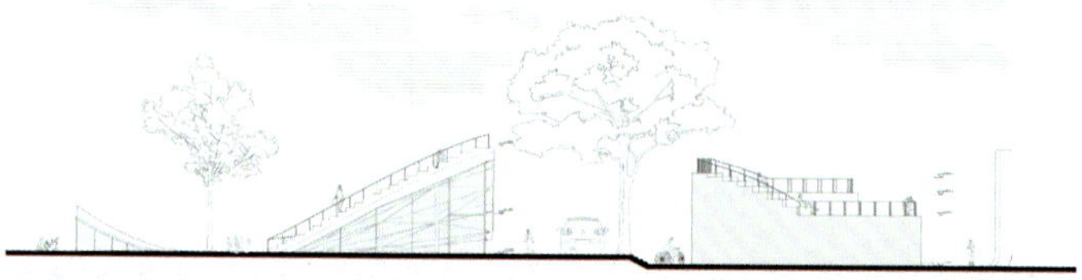

Facade

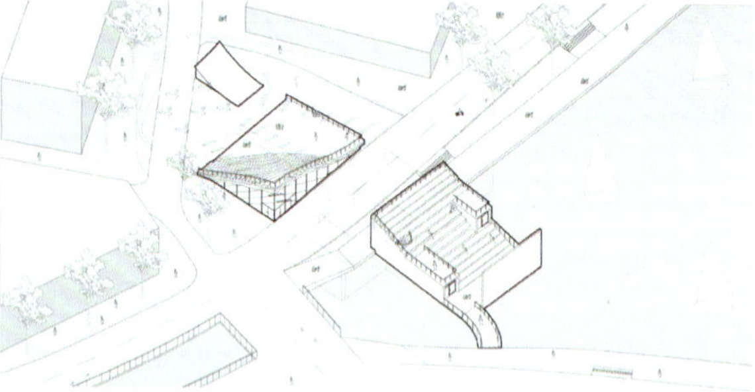

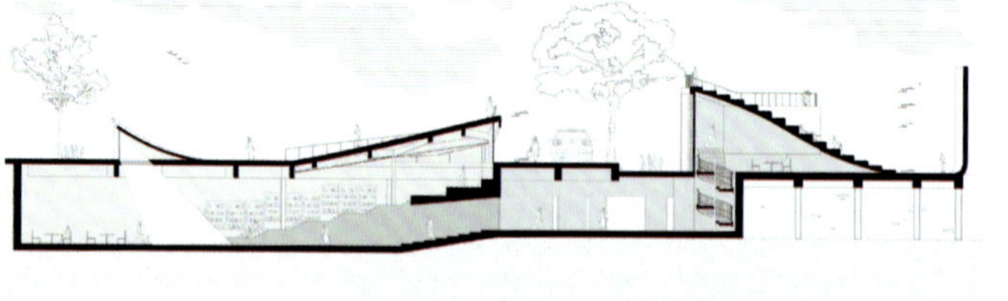

Section

■ 三等奖 Third Prize

润物“隙”无声

Silent and Soft, Space Moistens Everything

学校名称 University (College) Name
北京建筑大学，中国
Beijing University of Civil Engineering and Architecture, China

指导教师 Supervisor(s)
王　韬　Wang Tao

参赛学生 Participant(s)
李元亮　Li Yuanliang　　肖克强　Xiao Keqiang
夏　田　Xia Tian　　薛梓彬　Xue Zibin

设计说明 Design Description

以“润物细无声”为概念，以未来图书馆作为触媒，不破坏城区的肌理，以潜移默化的影响改变老城区功能单一等问题，同时希望把老城区缺乏活力的空间植入阅览室来逐渐改变人们阅读的习惯。通过把完整的图书馆变成不同功能区，分别植入老城中，虽然形式分离了，但功能仍是一个完整的图书馆。阅览室的藏书以及其他功能的分离，可大大增加布置的灵活性，因此阅览空间可以与其他功能组合置入老城，丰富老城的空间与功能，以多种方式激活老城区。

With the concept of “Silent and soft, it moistens everything”, and using the library of the future as a catalyst, we hope to change the single function of the old city without destroying the texture of the city, and to gradually change people’s reading habits by implanting a reading room in the negative space of the old city. Although the form is separated, the function is still a complete library, and the separation of the reading room from the book collection and other functions greatly increases the flexibility of the layout, so that the reading space can be combined with other functions and placed in the Old City, enriching the space and functions of the Old City and activating the Old City in many ways.

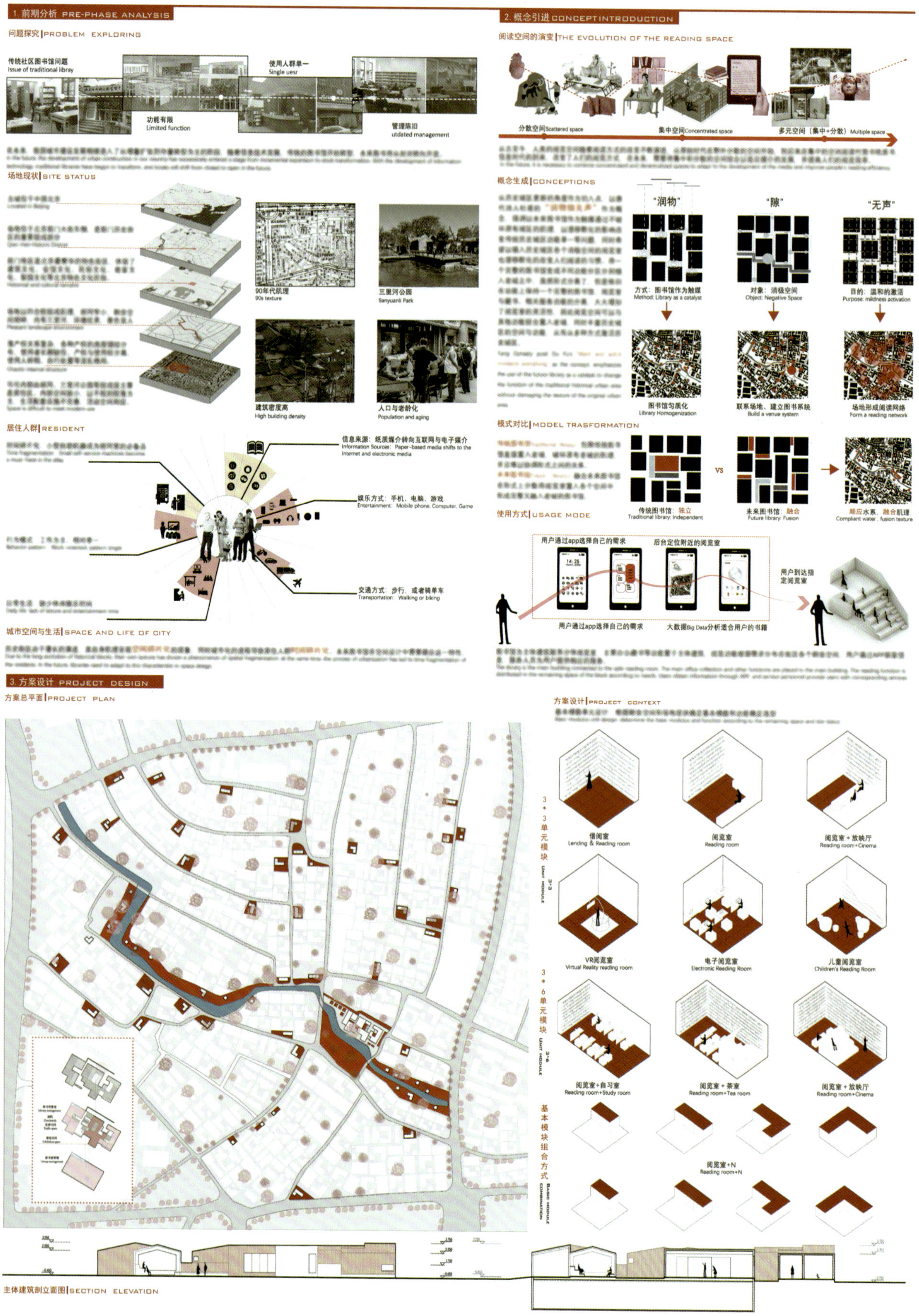
1. 前期分析 PRE-PHASE ANALYSIS
问题探究 | PROBLEM EXPLORING
传统社区图书馆问题
Issue of traditional library
使用人群单一
Single uesr
功能有限
Limited function
管理陈旧
utdated management
场地现状 | SITE STATUS
90年代肌理
90s texture
三里河公园
Sanyuanli Park
建筑密度高
High building density
人口与老龄化
Population and aging
居住人群 | RESIDENT
信息来源：纸质媒介转向互联网与电子媒介
Information Sources: Paper-based media shifts to the Internet and electronic media
娱乐方式：手机、电脑、游戏
Entertainment: Mobile phone, Computer, Game
交通方式：步行、或者骑单车
Transportation: Walking or biking
城市空间与生活 | SPACE AND LIFE OF CITY
2. 概念引进 CONCEPT INTRODUCTION
阅读空间的演变 | THE EVOLUTION OF THE READING SPACE
分散空间 Scattered space
集中空间 Concentrated space
多元空间（集中+分散）Multiple space
概念生成 | CONCEPTIONS
"润物"
"隙"
"无声"
方式：图书馆作为触媒
Method: Library as a catalyst
对象：消极空间
Object: Negative Space
目的：温和的激活
Purpose: mildness activation
图书馆匀质化
Library Homogenization
联系场地、建立图书系统
Build a venue system
场地形成阅读网络
Form a reading network
模式对比 | MODEL TRASFORMATION
传统图书馆：独立
Traditional library: Independent
VS
未来图书馆：融合
Future library: Fusion
顺应水系、融合肌理
Compliant water, fusion texture
使用方式 | USAGE MODE
用户通过app选择自己的需求
后台定位附近的阅览室
用户到达指定阅览室
用户通过app选择自己的需求
大数据Big Data分析适合用户的书籍
3. 方案设计 PROJECT DESIGN
方案总平面 | PROJECT PLAN
方案设计 | PROJECT CONTEXT
3*3单元模块 UNIT MODULE 3*3
借阅室
Lending & Reading room
阅览室
Reading room
阅览室+放映厅
Reading room+Cinema
VR阅览室
Virtual Reality reading room
电子阅览室
Electronic Reading Room
儿童阅览室
Children's Reading Room
3*6单元模块 UNIT MODULE 3*6
阅览室+自习室
Reading room+Study room
阅览室+茶室
Reading room+Tea room
阅览室+放映厅
Reading room+Cinema
基本模块组合方式 BASIC MODULE COMBINATION
阅览室+N
Reading room+N
主体建筑剖立面图 | SECTION ELEVATION

1. 选型分析 DESIGN SELECTION

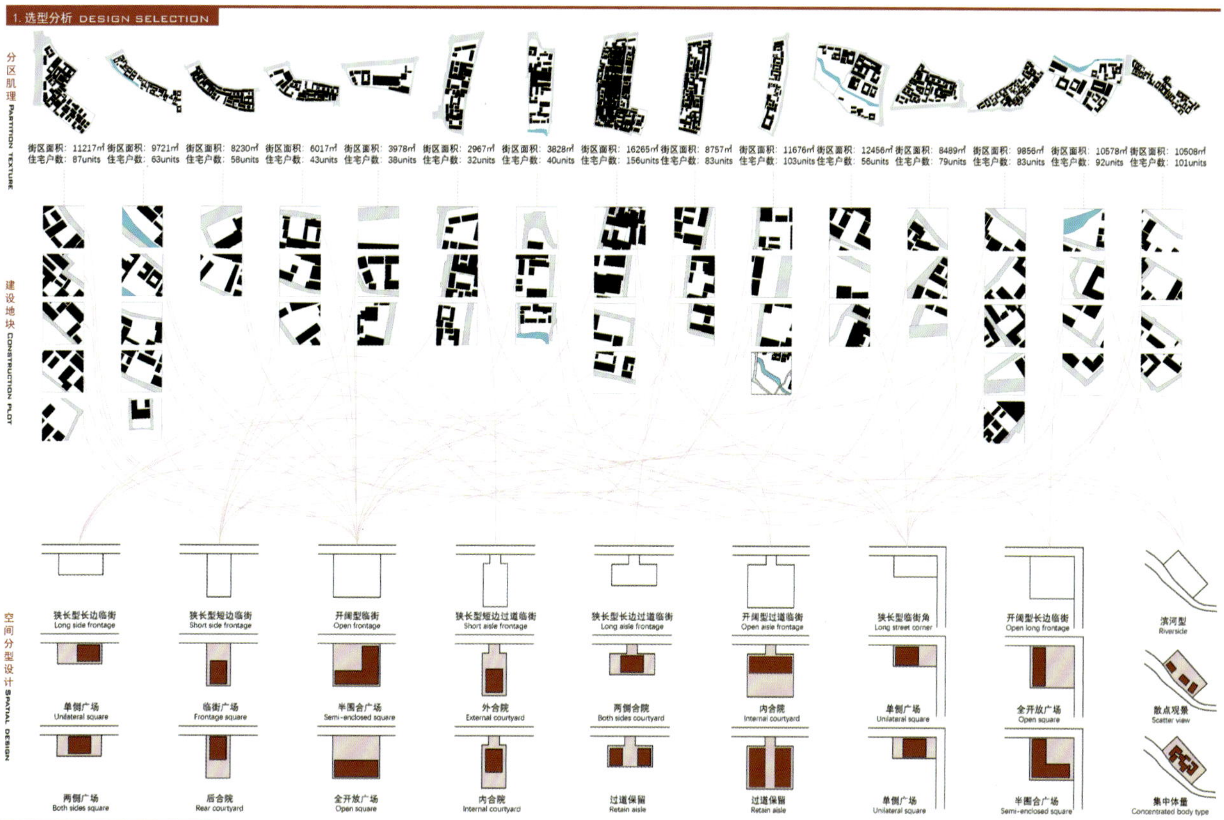

2. 鸟瞰图 AERIAL VIEW

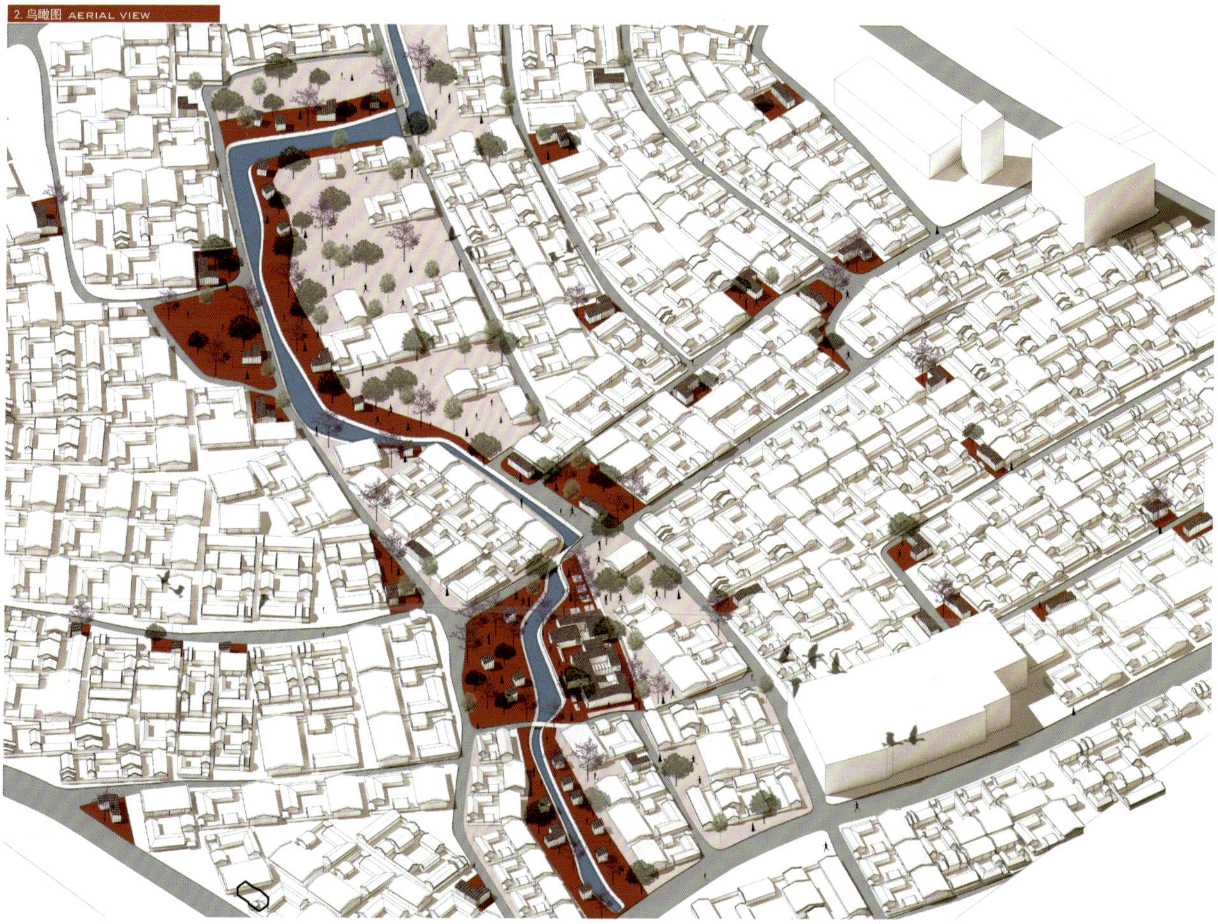

■ 三等奖 Third Prize

图书馆 +

Library Plus

学校名称 University (College) Name
长春建筑学院，中国
Changchun University of Architecture and Civil Engineering, China

指导教师 Supervisor(s)
张 蕾 Zhang Lei 滕佳佳 Teng Jiajia

参赛学生 Participant(s)
王儒钧 Wang Rujun 战 莹 Zhan Ying
陈小雨 Chen Xiaoyu 段 瑜 Duan Yu
葛志伟 Ge Zhiwei

设计说明 Design Description

进入信息时代以来，多元文化视角成为必然选择。图书馆作为信息的载体之一，曾极大地促进了信息的传播。但由于功能的单一性，逐渐失去了往日的优势。“图书馆 +”作为历史城区中的未来图书馆，连接历史和未来，探索新一代图书馆模式，以多元功能取代单一功能，以信息的传播与交流空间代替信息处理与查寻，以一切信息活动载体的定位代替图书阅读载体定位。同时“图书馆 +”作为一个系统，将与城市、自然、文脉等进行多角度的对话。

Since the dawn of the information era, multi-cultural perspective has become an inevitable choice. As one of the information carrier, the library has greatly promoted the dissemination of information. However, due to the singleness of its functions, it has gradually lost its advantages of the past. As the future library in the historical city, “Library Plus” connects the history with the future, and explores the mode of a new generation of library to replace single function with multiple functions, replace information processing and search with information dissemination and exchange space, and replace the positioning of book reading carrier with the positioning of all information activity carriers. Meanwhile, “Library Plus”, as a system, will conduct multi-angle dialogues with cities, nature and context.

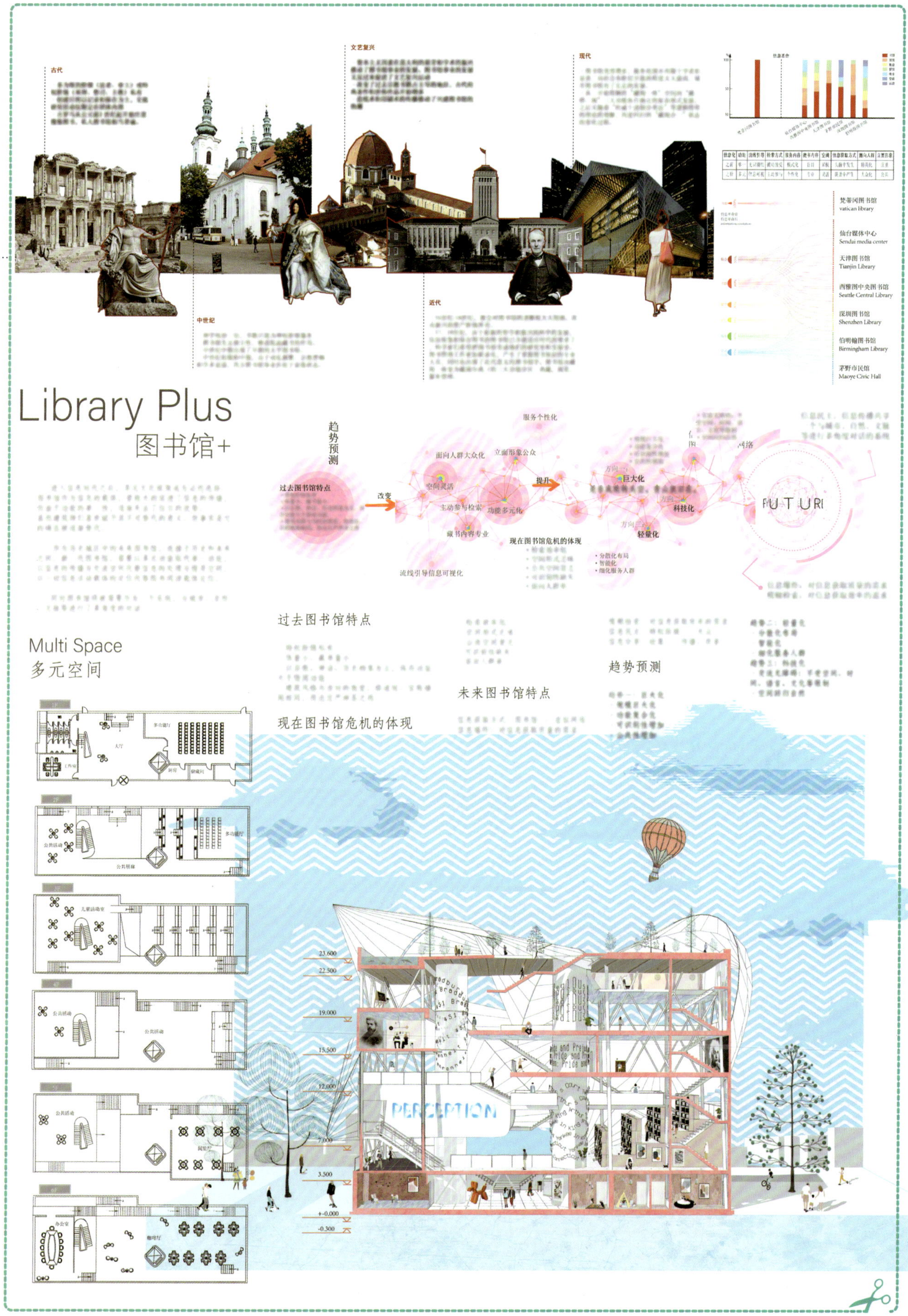
古代
文艺复兴
现代
中世纪
近代
梵蒂冈图书馆
vatican library
仙台媒体中心
Sendai media center
天津图书馆
Tianjin Library
西雅图中央图书馆
Seattle Central Library
深圳图书馆
Shenzhen Library
伯明翰图书馆
Birmingham Library
茅野市民馆
Maoye Civic Hall
Library Plus
图书馆+
趋势预测
服务个性化
面向人群大众化
立面形象公众
过去图书馆特点
改变
空间灵活
主动参与检索
功能多元化
提升
巨大化
科技化
轻量化
藏书内容专业
现在图书馆危机的体现
流线引导信息可视化
FUTURE
过去图书馆特点
未来图书馆特点
现在图书馆危机的体现
趋势预测
Multi Space
多元空间
23.600
22.500
19.000
15.500
12.000
7.000
3.500
+-0.000
-0.300
PERCEPTION

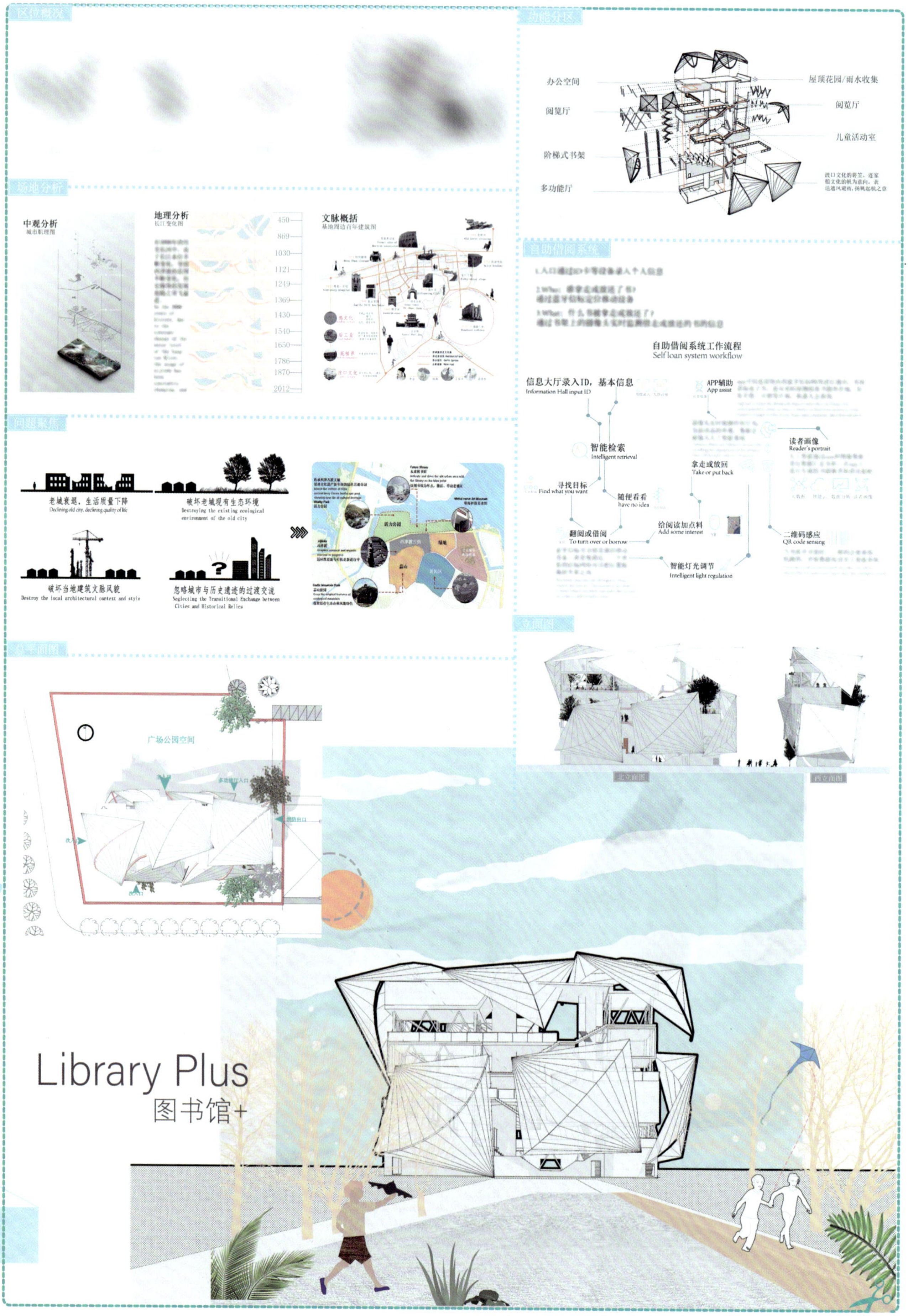
区位概况
功能分区
办公空间
阅览厅
阶梯式书架
多功能厅
屋顶花园/雨水收集
阅览厅
儿童活动室
场地分析
中观分析
城市肌理图
地理分析
长江变化图
450
869
1030
1121
1249
1369
1430
1510
1650
1786
1870
2012
文脉概括
基地周边百年建筑图
自助借阅系统
自助借阅系统工作流程
Self loan system workflow
信息大厅录入ID，基本信息
Information Hall input ID
APP辅助
App assist
智能检索
Intelligent retrieval
读者画像
Reader's portrait
拿走或放回
Take or put back
寻找目标
Find what you want
随便看看
have no idea
翻阅或借阅
To turn over or borrow
给阅读加点料
Add some interest
二维码感应
QR code sensing
智能灯光调节
Intelligent light regulation
问题聚焦
老城衰退，生活质量下降
Declining old city, declining quality of life
破坏老城现有生态环境
Destroying the existing ecological environment of the old city
破坏当地建筑文脉风貌
Destroy the local architectural context and style
忽略城市与历史遗迹的过渡交流
Neglecting the Transitional Exchange between Cities and Historical Relics
总平面图
广场公园空间
立面图
Library Plus
图书馆+

■ 三等奖　Third Prize

川河未来·新陈代谢

Future of Chuanhe • Metabolism River

学校名称 University (College) Name
吉林建筑大学，中国
Jilin Jianzhu University, China

指导教师 Supervisor(s)
张广平　Zhang Guangping　　常　悦　Chang Yue

参赛学生 Participant(s)
陈梦薇　Chen Mengwei　　郭文博　Guo Wenbo
韩春苗　Han Chunmiao　　刘伯琪　Liu Boqi
吴文杰　Wu Wenjie

设计说明 Design Description

本设计项目选址在厦门新老城区交汇处临海建设图书馆，以此过渡新老城矛盾冲突，为老城区注入新的活力。建筑迎合未来海平面上升的危机状况，预先设计符合未来生长模式的建筑形式；结合 6G 数字孪生时代技术，打造超前建筑空间体验；核心筒的结构体系与装配式建筑施工结合在一起，加快施工速度，灵活的施工形式有效地预防台风、海啸和其他危机；单元体块也可上下移动变换，以此实现建筑结构与特性的多样化。

The design involves building a library near the sea at the intersection of Xiamen's new and old towns as a transition to solve the conflicts between the two sides and inject new vitality into the old. In view of the crisis of rising sea levels in the future, an architectural form that conforms to the future growth model is designed; combined with the technology of the 6G digital twin era, an advanced architectural space is created to provide immersive reading experience; the core tube structure system is combined with the prefabricated building construction to speed up the construction progress, and the flexible construction form can effectively prevent damage of typhoons, tsunamis and other crises; the unit blocks can also be moved up and down to diversify the building's layout and functions.

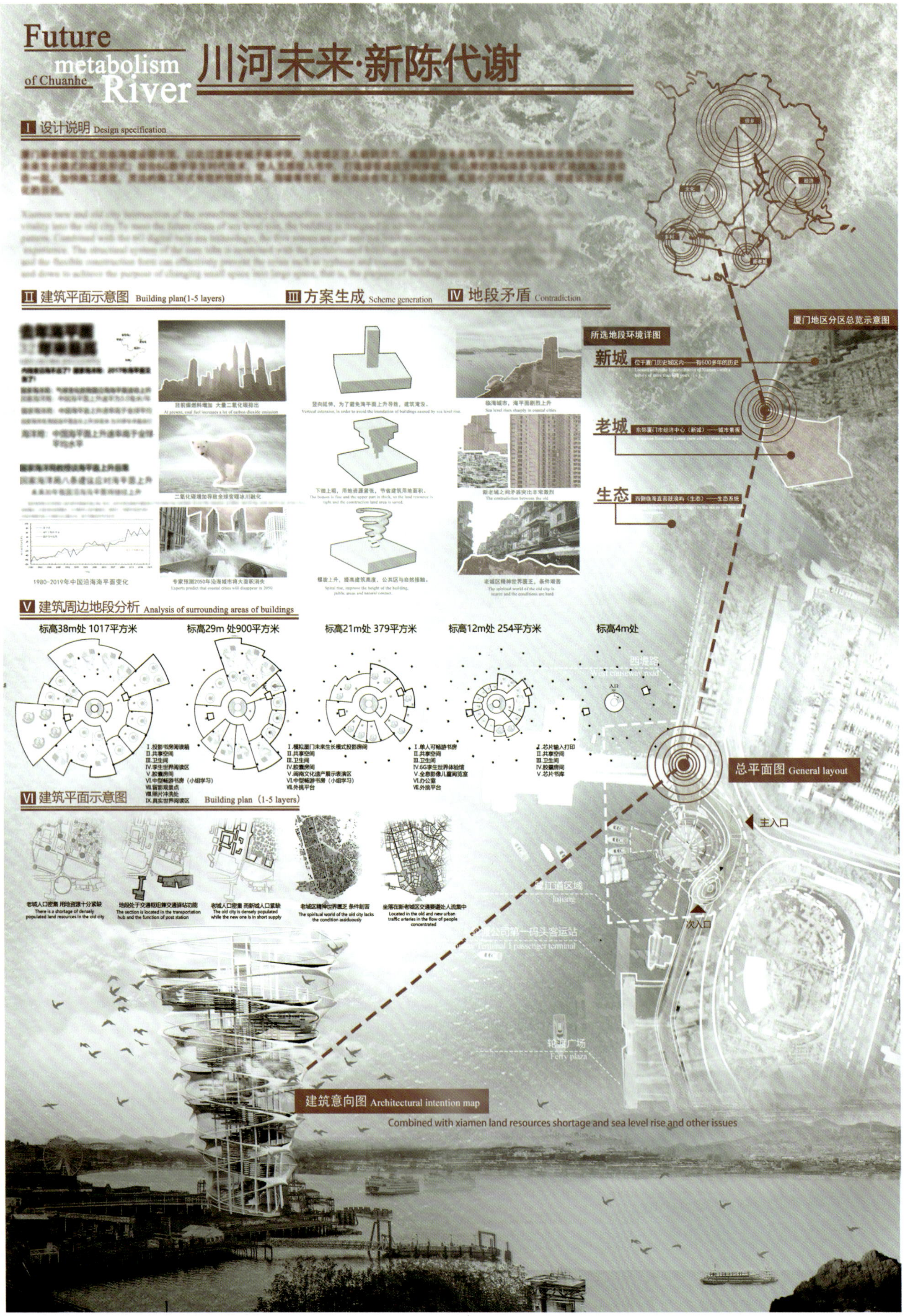
Future metabolism of Chuanhe River
川河未来·新陈代谢
Ⅰ 设计说明 Design specification
Ⅱ 建筑平面示意图 Building plan(1-5 layers)
Ⅲ 方案生成 Scheme generation
Ⅳ 地段矛盾 Contradiction
目前煤燃料增加 大量二氧化碳排出
二氧化碳增加导致全球变暖冰川融化
专家预测2050年沿海城市将大面积消失
1980-2019年中国沿海海平面变化
竖向延伸，为了避免海平面上升导致，建筑淹没。
下狭上粗，用地资源紧张，节省建筑用地面积。
螺旋上升，提高建筑高度，公共区与自然接触。
临海城市，海平面剧烈上升
新老城之间矛盾突出非常激烈
老城区精神世界匮乏，条件堪忧
所选地段环境详图
厦门地区分区总览示意图
新城
老城
生态
Ⅴ 建筑周边地段分析 Analysis of surrounding areas of buildings
标高38m处 1017平方米
标高29m 处900平方米
标高21m处 379平方米
标高12m处 254平方米
标高4m处
西堤路
West causeway road
入口
Ⅵ 建筑平面示意图 Building plan (1-5 layers)
老城人口密集 用地资源十分紧缺
There is a shortage of densely populated land resources in the old city
地段处于交通枢纽兼交通驿站功能
The section is located in the transportation hub and the function of post station
老城人口密集 而新城人口紧缺
The old city is densely populated while the new one is in short supply
老城区精神世界匮乏 条件刻苦
The spiritual world of the old city lacks the condition assiduously
坐落在新老城区交通要道处人流集中
Located in the old and new urban traffic arteries in the flow of people concentrated
总平面图 General layout
主入口
次入口
鹭江道区域
Lujiang
轮渡广场
Ferry plaza
建筑意向图 Architectural intention map
Combined with xiamen land resources shortage and sea level rise and other issues

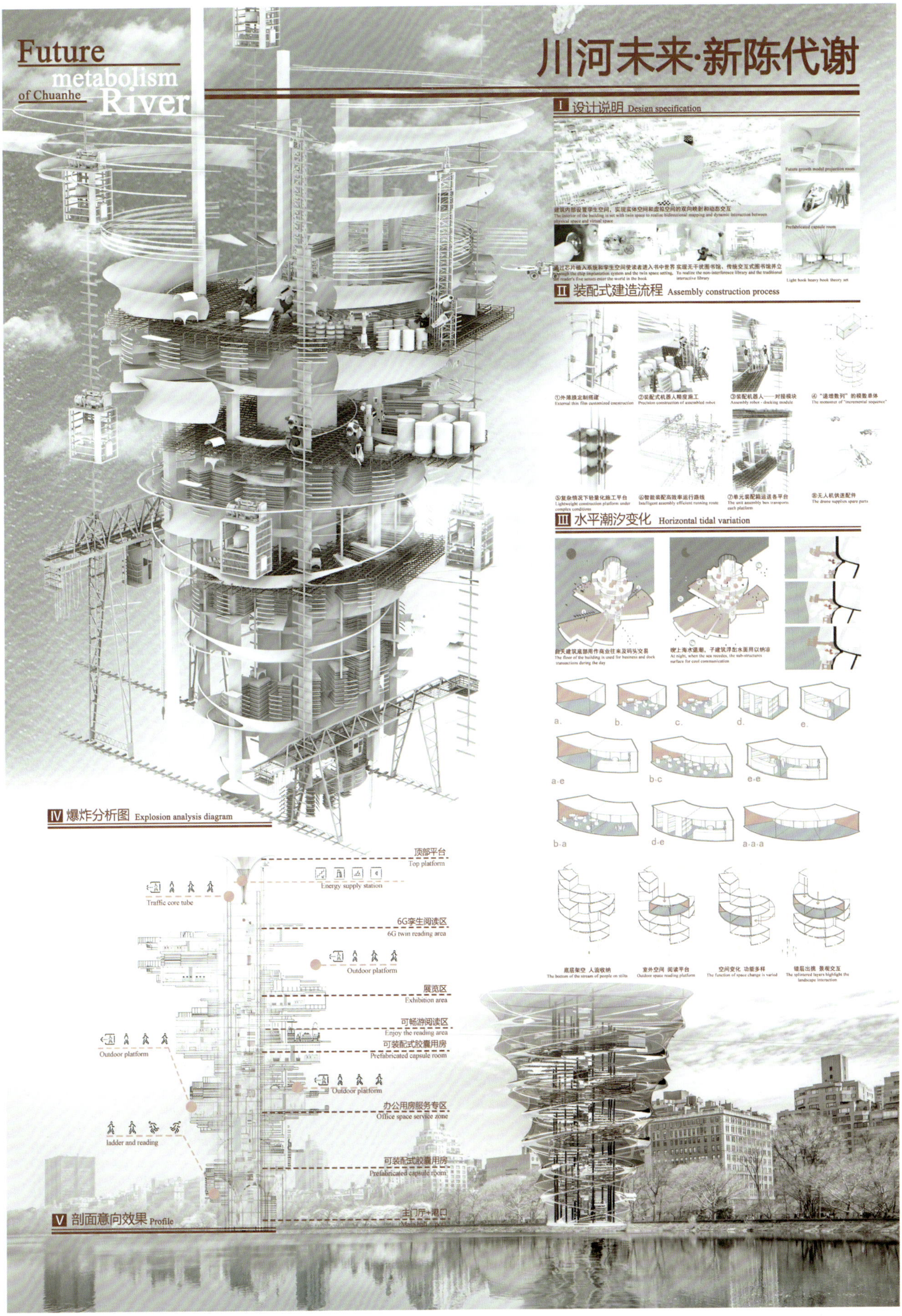
Future
metabolism
of Chuanhe
River
川河未来·新陈代谢
I 设计说明 Design specification
建筑内部设置孪生空间，实现实体空间和虚拟空间的双向映射和动态交互
The interior of the building is set with twin space to realize bidirectional mapping and dynamic interaction between physical space and virtual space
Future growth model projection room
Prefabricated capsule room
通过芯片植入系统和孪生空间使读者进入书中世界
Through the chip implantation system and the twin space setting, the reader's five senses enter the world in the book
实现无干扰图书馆、传统交互式图书馆并立
To realize the non-interference library and the traditional interactive library
Light book heavy book theory set
II 装配式建造流程 Assembly construction process
①外薄膜定制搭建
External thin film customized construction
②装配式机器人精度施工
Precision construction of assembled robot
③装配机器人——对接模块
Assembly robot - docking module
④"递增数列"的模数单体
The monomer of "incremental sequence"
⑤复杂情况下轻量化施工平台
Lightweight construction platform under complex conditions
⑥智能装配高效率运行路线
Intelligent assembly efficient running route
⑦单元装配箱运送各平台
The unit assembly box transports each platform
⑧无人机供送配件
The drone supplies spare parts
III 水平潮汐变化 Horizontal tidal variation
白天建筑底部用作商业往来及码头交易
The floor of the building is used for business and dock transactions during the day
晚上海水退潮，子建筑浮出水面用以纳凉
At night, when the sea recedes, the sub-structures surface for cool communication
a.
b.
c.
d.
e.
a-e
b-c
e-e
b-a
d-e
a-a-a
底层架空 人流收纳
The bottom of the stream of people on stilts
室外空间 阅读平台
Outdoor space reading platform
空间变化 功能多样
The function of space change is varied
错层出挑 景观交互
The splintered layers highlight the landscape interaction
IV 爆炸分析图 Explosion analysis diagram
顶部平台
Top platform
Energy supply station
Traffic core tube
6G孪生阅读区
6G twin reading area
Outdoor platform
展览区
Exhibition area
可畅游阅读区
Enjoy the reading area
可装配式胶囊用房
Prefabricated capsule room
Outdoor platform
Outdoor platform
办公用房服务专区
Office space service zone
ladder and reading
可装配式胶囊用房
Prefabricated capsule room
主门厅+港口
Main hall + port
V 剖面意向效果 Profile

■ 三等奖 Third Prize

地下空间

The Chamber Under

学校名称 University (College) Name
加德满都大学，尼泊尔
Kathmandu University, Nepal

指导教师 Supervisor(s)
Aakash Bhochhibhoya

参赛学生 Participant(s)
Sudipti Giri
Ilam Shrestha
Sansu Shrestha
Nandita Shrestha
Bidhya Gupta

设计说明 Design Description

历史城区的未来图书馆。

“坚守当下和当地特色，连接过去与未来。”

——詹姆斯·乔伊斯

以“地下空间＂为理念，在历史悠久的城区设计一座未来化的图书馆，代表着将未来空间与过去相连接的概念。

图书馆的选址在众神之地——加德满都，根据当地文化，人们认为水域是神圣而有灵性的。但讽刺的是，他们也把水域当作垃圾场。

在尼泊尔加德满都郊区的基尔蒂普尔古镇，山谷中有一个名为陶达哈（Taudaha）的小湖还相对干净。据信，那里是蛇王的居所，其背后的神话故事令这片湖泊免于成为废墟。

实际上，人们也留下了一块不受干扰的土地，为真实世界或神话里的蛇留下了一片空间，也给了候鸟和水生生物一个多年的栖息空间。

因此，图书馆将通过未来感的技术来传达过去的神话故事，如带有全息投影的地下隧道、虚拟现实、透明玻璃屏幕将向游客展示水下蛇王的生活方式。正如历史是这个城镇最重要的一部分一样，从图书馆外面几乎不会看到上述这些内容，但它们是图书馆最重要的一部分。

该项目的理念围绕着下述概念的共存：

- 过去与未来
- 建筑与大自然
- 不同性质的用户

因此，不考虑差异性，为每个人创造“地下空间”——这是人们在未来的需要。在这个图书馆，游客能够从视觉上了解当地的历史和生态圈。

Future Library in a Historical Urban Area.

"Hold to the now, the here, through which all future plunges to the past"

— James Joyce

Designing a future library in a historical urban area instigated the notion of "The Chamber Under" as a future space to connect with the past.

The site chosen is in Kathmandu, the land of the Gods, with a culture that considers water bodies to be holy and spiritual, yet ironically treating them as dumping sites.

A small lake, Taudaha, in the ancient town of Kirtipur, the outskirts of Kathmandu, Nepal is a comparatively clean water body remaining in the valley. Believed to be the abode of serpent kings, the mythology that backs it up has saved it from ruins.

The fact that people have left a place untouched to allow snakes- real and mythological their space, gives hope that migratory birds and aquatic life will still have a space to rest for many years still.

Thus, futuristic technology is used in the library to convey legends of the past, like the underground tunnel with hologram projections, virtual reality, transparent glass screens to show visitors the lifestyle of serpent kings underwater. This portion is barely seen from the outside, yet is the most important part of the library, much like history is the most important part of the town.

The concept of this project revolves around coexistence of

- The Past and the Future
- Architecture and Nature
- Users of different nature

Thus, making "The Chamber Under" space for everyone, irrespective of their differences, the need of the future. It is cultivated as a place where visitors get to visually learn, among other things, about the history and ecosphere of the place.

SITE BRIEF

SITE ANALYSIS

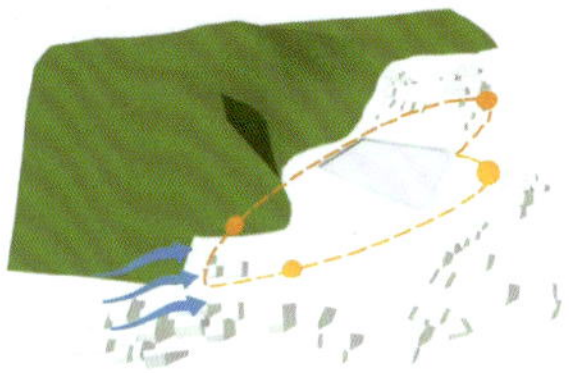

CONCEPTUAL DEVELOPMENT

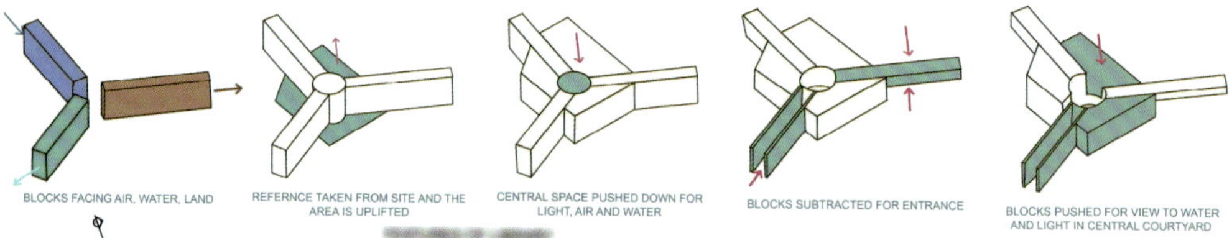

NORTHWEST VIEW OF THE BUILDING

AERIAL VIEW OF THE BUILDING

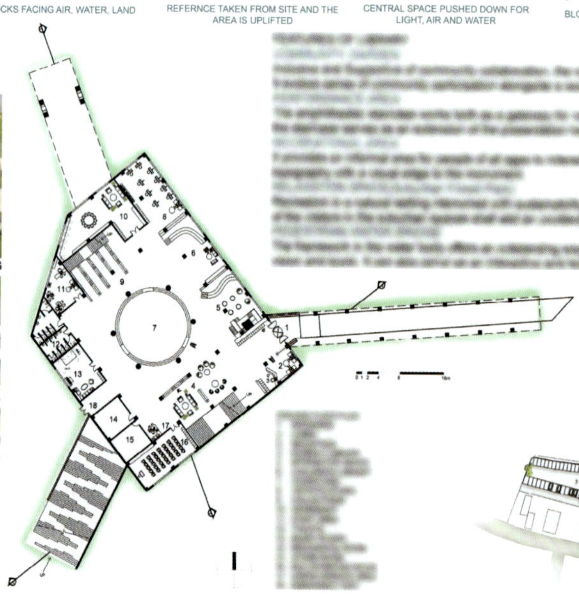

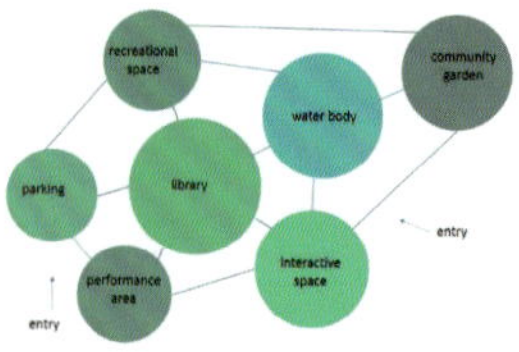

ZONING OF SPACES

EAST ELEVATION

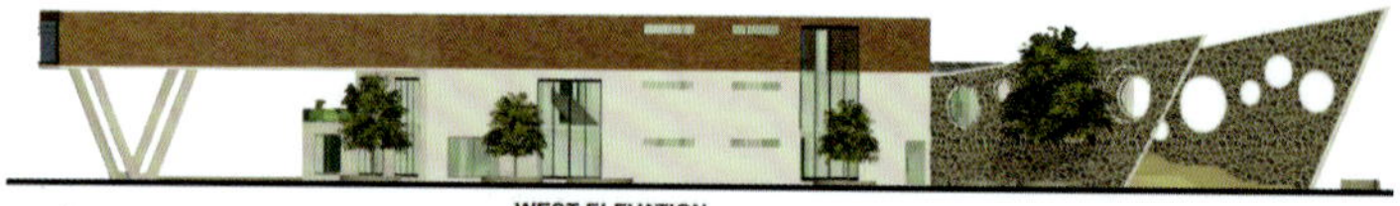

WEST ELEVATION

PUSTAKAGAR ...the library of future

INTERIOR VIEW OF THE GREAT READING ROOM

SOUTHWEST VIEW OF BUILDING

ROOF TOP TERRACE VIEW

SOUTHEAST VIEW OF BUILDING

INTERIOR VIEW OF THE CENTRAL COURTYARD

EXTERIOR VIEW OF THE ENTRY WAY

INTERIOR VIEW OF MEDITATION SPACE

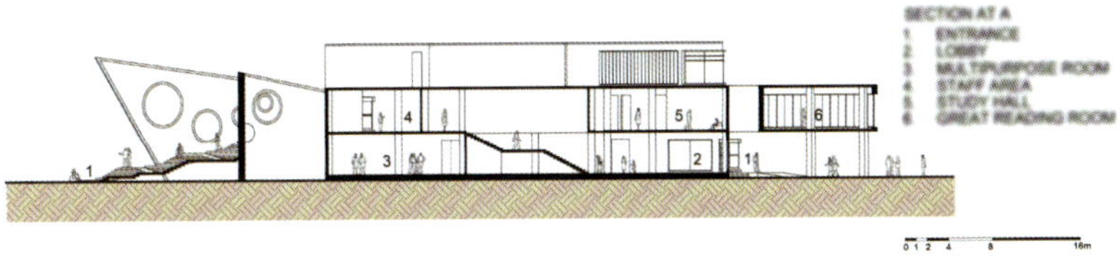

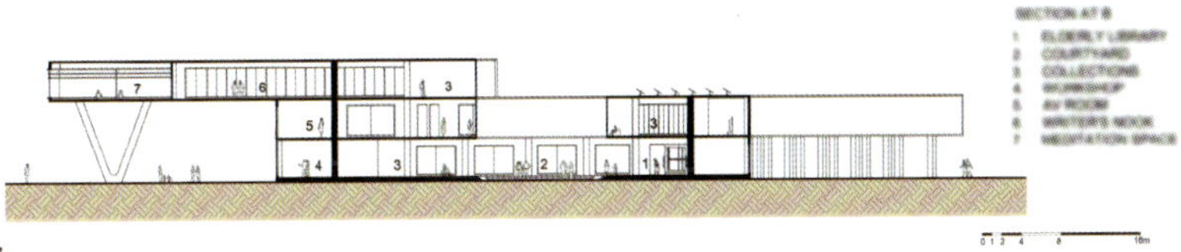

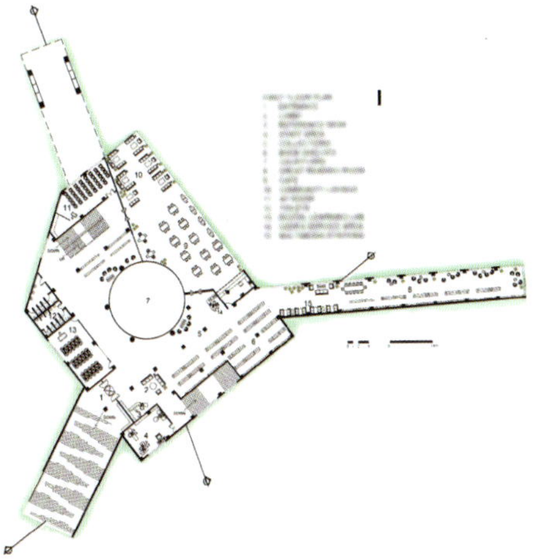

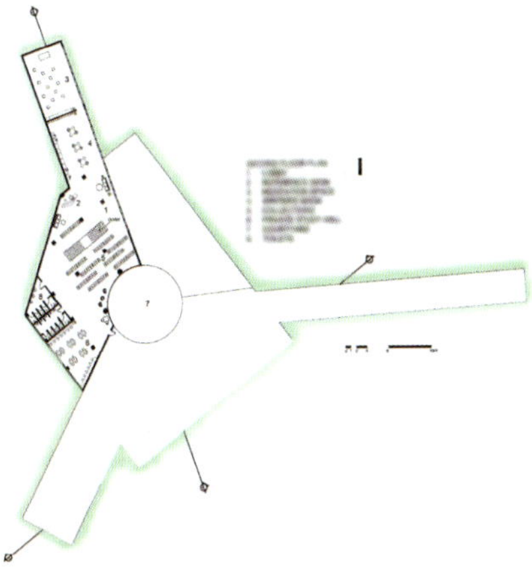

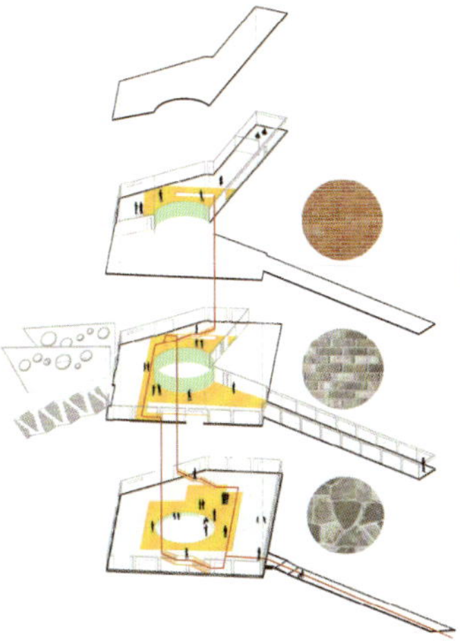

EXPLODED AXONOMETRIC VIEW

NORTH ELEVATION

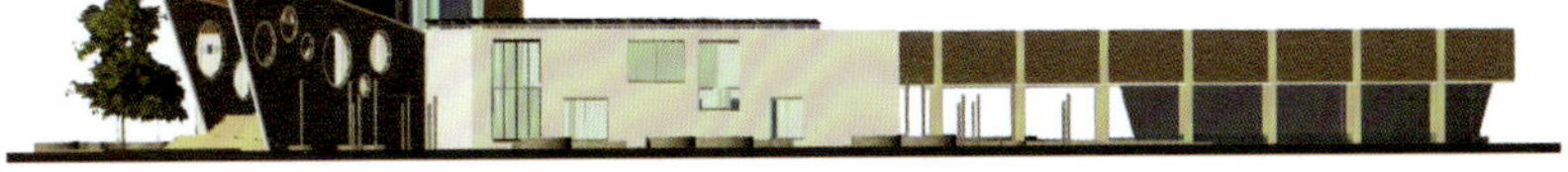

SOUTH ELEVATION

PUSTAKAGAR ...the library of future

■ 三等奖 Third Prize

一生一景

One Life, One Scene

学校名称 University (College) Name
莫斯科国立建筑大学，俄罗斯
Moscow State University of Civil Engineering, Russia

指导教师 Supervisor(s)
Pushkina Natalya

参赛学生 Participant(s)
Berseneva Anastasiia

设计说明 Design Description

改造位置合适的废弃工厂，将其改造为可开展各类活动的新图书馆。图书馆改造的完整历程，包括为什么选址在这里、将来计划如何开发等，都用图纸展示出来。以下是设计信息：图书馆由两座状况良好的老建筑改建而成。从节约资源和可持续发展的角度来看，并没有摧毁重建它们的必要，当然，不保留此外的附属建筑。因为废弃厂房的面积相当大，所以空间规划上不需要过于精打细算。我设计的图书馆是一个各种功能一应俱全的综合体，会在原有梁柱的基础上，新建图书馆的立面。我计划使用透光混凝土等新型材料来建造该综合设施，另外，展厅和酒吧的外立面会采用非透明玻璃。这样，建筑在视觉上会集中于图书馆的部分，从而消除构图上的突兀。圆柱体会成为图书馆的核心，独特的造型总是引人注目且令人印象深刻。

Old abandoned factory in good location will become a new library with different activities inside. Every step why this territory of Moscow was choosen and how the area will be developed you will see on the posters. Here I want to put some design information. The library will be located in two old buildings that are in good condition. That is also sustainable approach as there is no need to destroy the one that can be renovated. Other buildings will be demolish. On there place there will be a small but functional complex. The area is quite big that's why we did not need to cut the space. In base of the library will remain a post-beam construction system with new facades. In the multifunctioning complex the new materials will be applied such as litracon. Facades made out of non transparent glass of exhibition hall, bar will make a view to the library more focused and will lite the composition. Cylinder will become the core of the area as a semantic dominant. That's why this rare form of building will also give the library a memorable and eye-catching look.

ONE LIFE - ONE SCENE
HISTORY OF THE PLACE
CONCEPT DIAGRAM

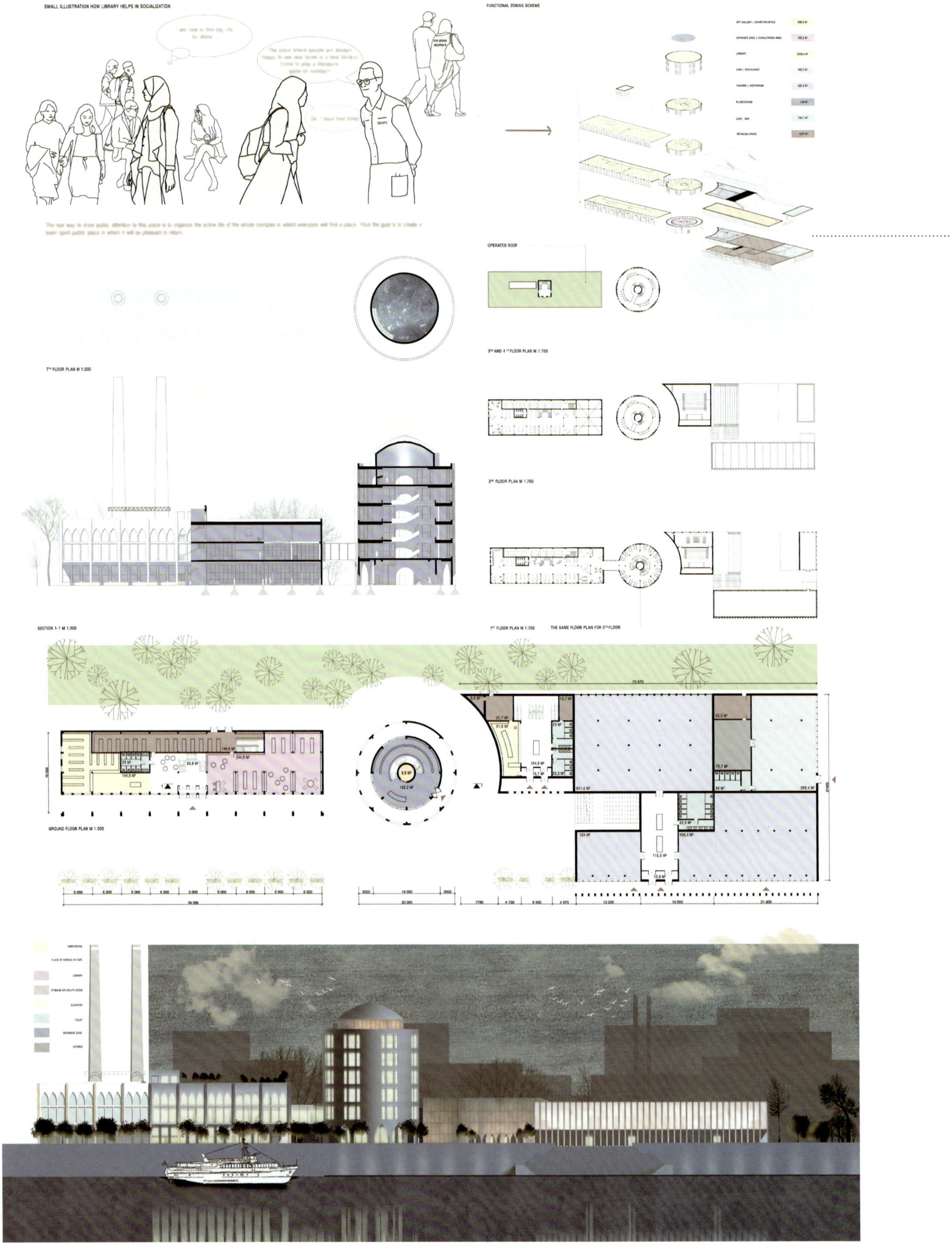
SMALL ILLUSTRATION HOW LIBRARY HELPS IN SOCIALIZATION
FUNCTIONAL ZONING SCHEME
OPERATED ROOF
3RD AND 4TH FLOOR PLAN M 1:700
7TH FLOOR PLAN M 1:300
2ND FLOOR PLAN M 1:700
SECTION 1-1 M 1:300
1ST FLOOR PLAN M 1:700
THE SAME FLOOR PLAN FOR 5TH FLOOR
GROUND FLOOR PLAN M 1:300

■ 三等奖　Third Prize

LibraryN +・茧生

LibraryN +・Jiansheng

学校名称　University (College) Name
山东建筑大学，中国
Shandong Jianzhu University, China

指导教师　Supervisor(s)
门艳红　Men Yanhong　　郑恒祥　Zheng Hengxiang

参赛学生　Participant(s)
刘相宇　Liu Xiangyu　　张聿柠　Zhang Yuning
王庆峰　Wang Qingfeng　　徐化超　Xu Huachao
安兴腾　An Xingteng

设计说明 Design Description

本方案选取济南市芙蓉街—百花洲老城区作为基地选择片区。依托于片区内的传统泉池位置与历史建筑的辐射范围，选择四块基地，以图书馆功能为主体，根据每个基地的不同特点，进行附加一个新功能的模式选择。新功能系符合基地周围居民生活所需和游客所需等多重意义下的产物，包括居住的旅店、休闲的茶室、聚会的酒吧、演出的戏台等等。我们通过斟酌，选择四个基地进行四个方案的设计（侧重于一个方案的表达），阐释一种对于老城区不同基地合体功能的处理模式。

The former urban area from Furong Street to Baihua Pool of Jinan Municipality is the site where the selected bases are located. These four bases are selected on the basis of the location of the traditional spring pool in the area and the influence range of historical buildings. The library functions as the main body, and an additional new function mode is provided according to the different characteristics of each base. The new function, as a result of tourists' diverse demands, can meet the needs of the residents around the base, for example, the demand for a hotel, a tea house, a bar for a party, a stage for a show. After careful consideration, we chose four bases for the design of four schemes (but focused on the expression of one scheme) to explain the function combination of different bases in the former urban area.

Library N+·茧生 I

设计说明
Design Description

建设范围
Build Range

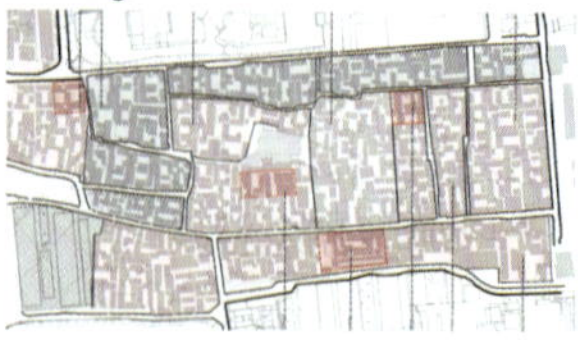

场地选择
Site Selection

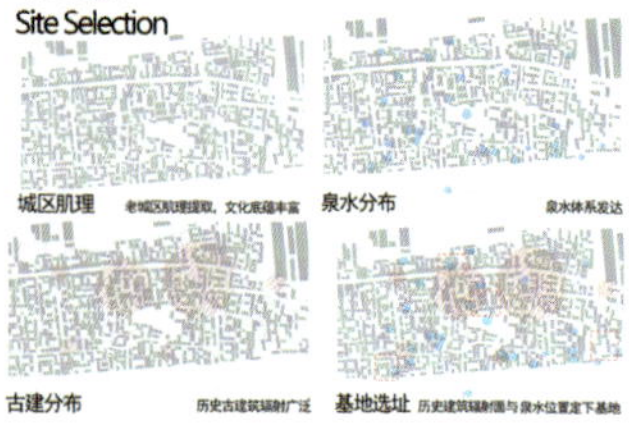

场地分析
Site Analysis

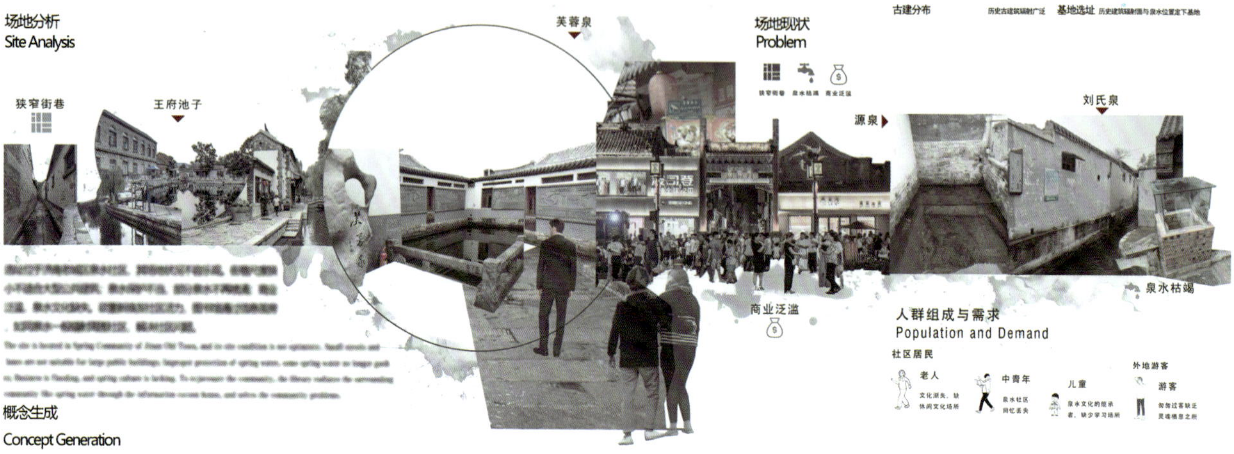

概念生成
Concept Generation

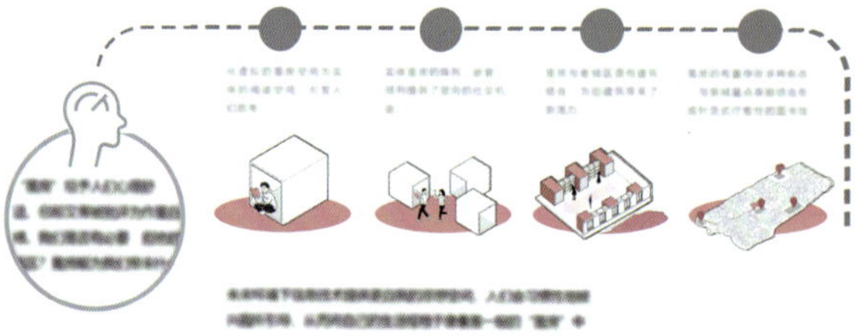

场景透视 Scene Perspective

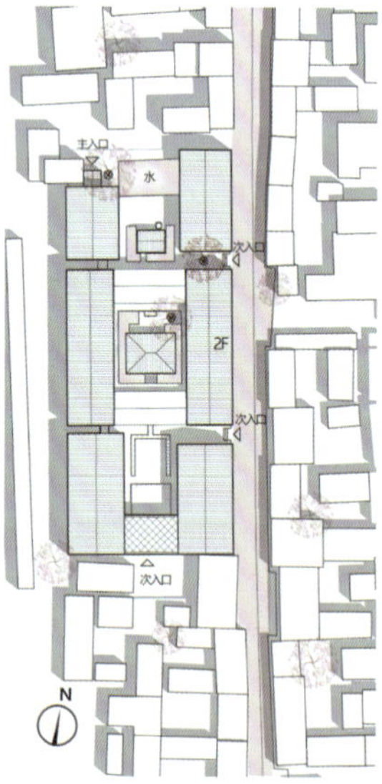

总平面图 Site 1:500

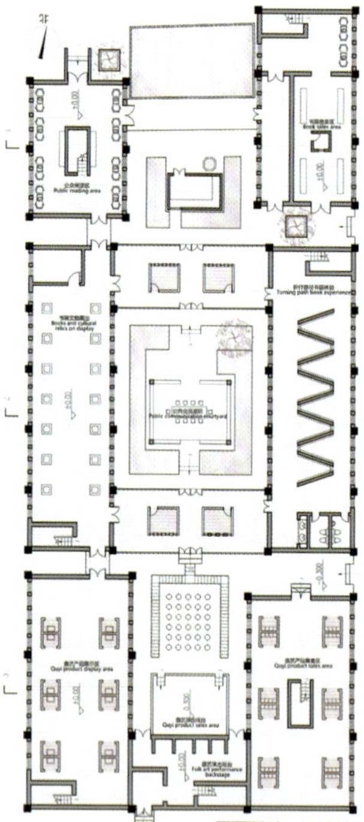

一层平面 Plan1 1:200

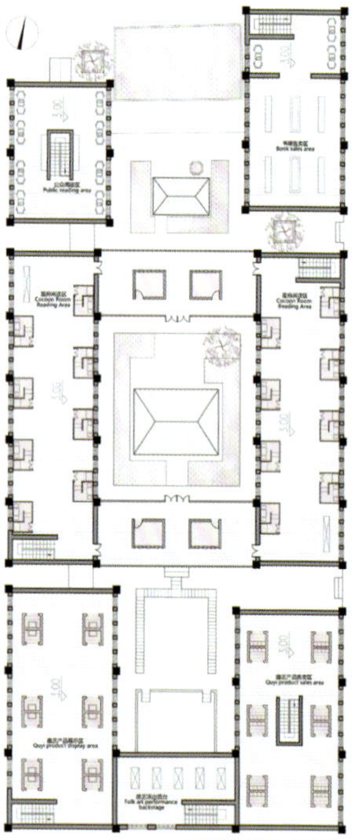

二层平面 Plan2 1:200

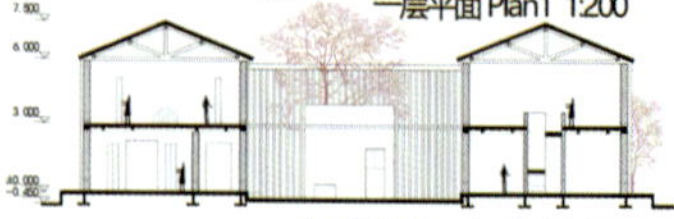

剖面图1-1 Section1-1 1:200

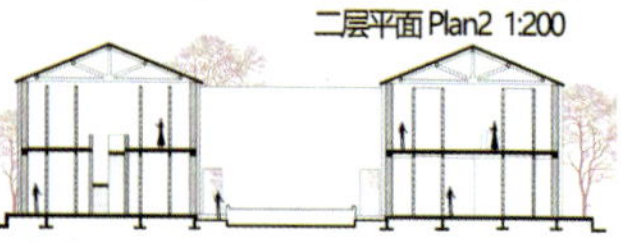

剖面图3-3 Section3-3 1:200

方案生成 Plan Generation

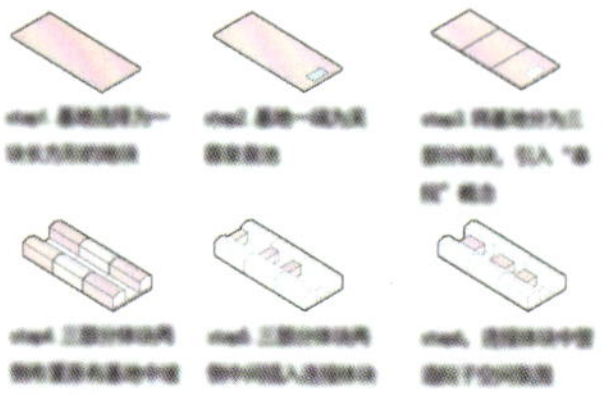

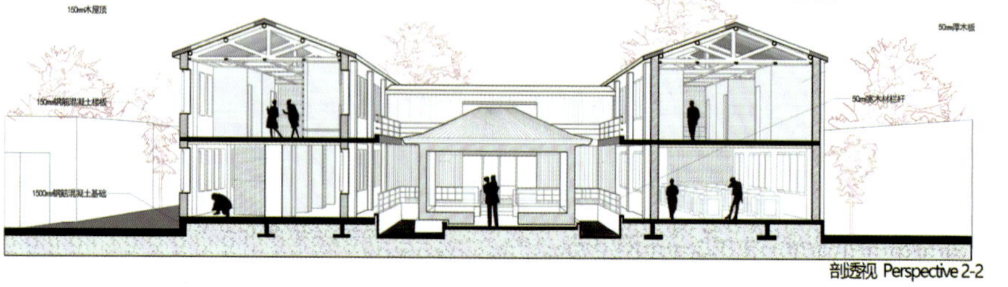

剖透视 Perspective 2-2

Library N+·茧生 II

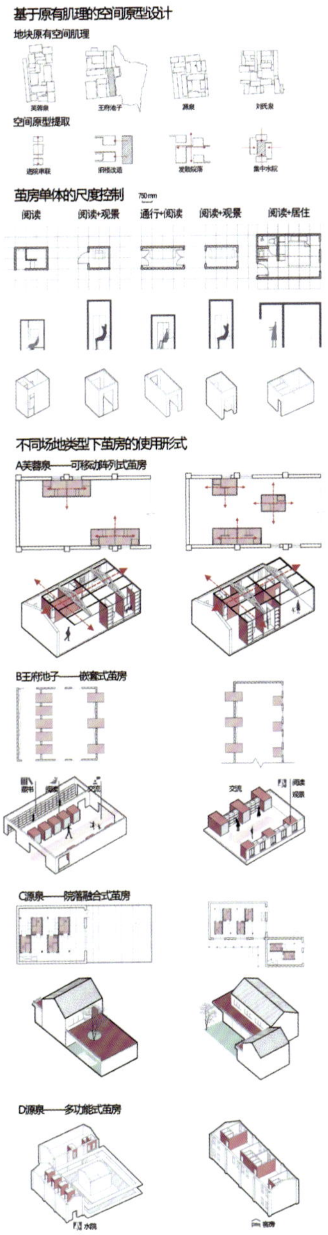

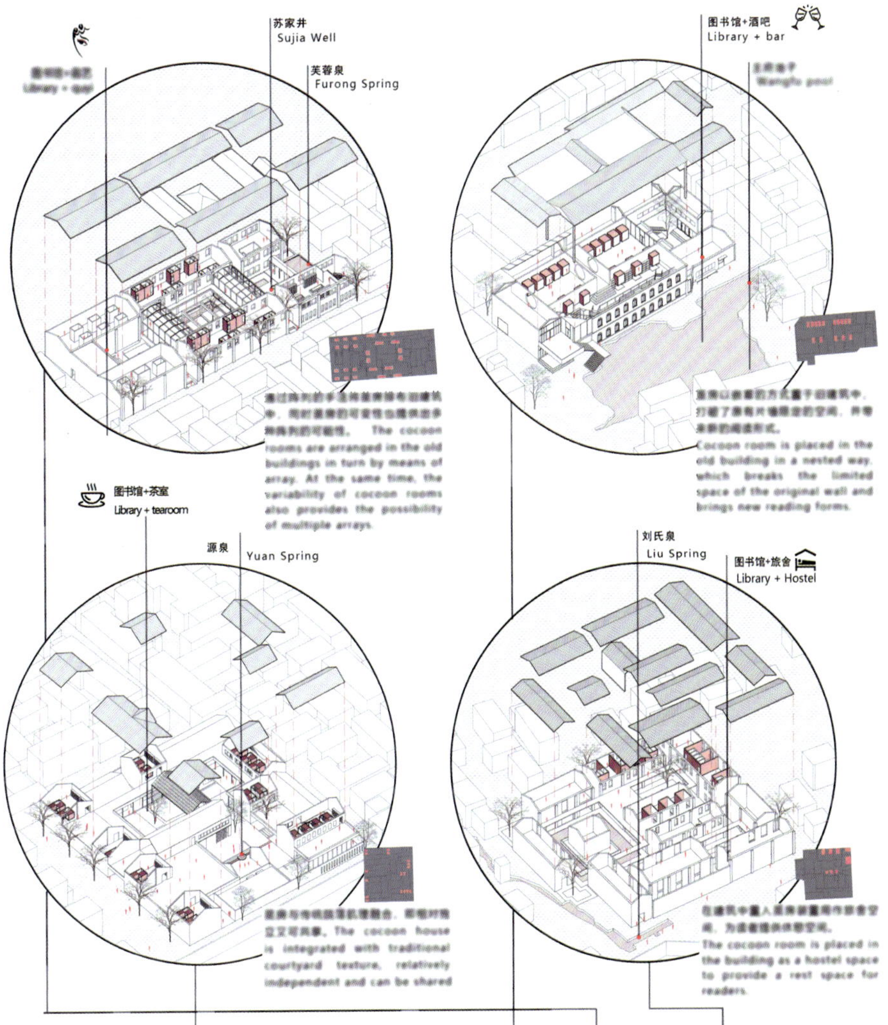

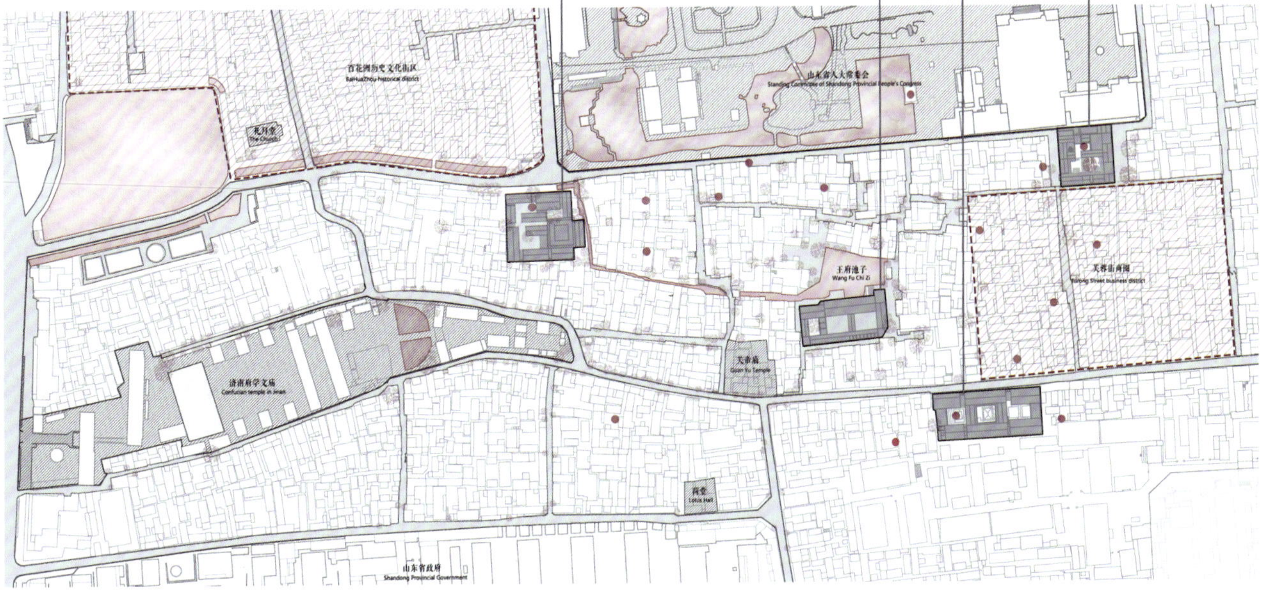

■ 三等奖 Frist Prize

喧社 · 静息

Tranquil Life within Vibrant Community

学校名称 University (College) Name
天津城建大学，中国
Tianjin Chengjian University, China

指导教师 Supervisor(s)
刘 辉 Liu Hui　　万 达 Wan Da

参赛学生 Participant(s)
卢 聪 Lu Cong　　马 龙 Ma Long

设计说明 Design Description

本设计从杨柳青古镇的现况出发，根据泰森多边形理论基于现实肌理，提取合院元素置入公共服务设施体块。在畅想未来极为完善的社区状态下，伴随着数字信息的影响，社区将成为一个功能丰富、信息纷杂的完整组团。快节奏生活下的居民逐渐向往纯粹的宁静。未来图书馆或将成为承载居民内心宁静寄托的场所。与此同时，结合本地文化创造出居民与游客的交互场所，并提出游客与居民相处的新模式，从而提升杨柳青的地区文化价值。

This design starts from the current situation of Yangliuqing ancient town, according to Tyson polygon theory, we extract courtyard elements based on realistic texture and put them into public service facilities block. With the influence of digital information, the community will become a complete cluster with rich functions and various information. Under the fast-paced life, the residents gradually yearn for the pure tranquility. In the future, the library will become the place to carry the residents' inner peace. At the same time, we combine with the local culture to create a place for interaction between residents and tourists, we also put forward a new model for tourists and residents to get along with each other, so as to enhance the regional cultural value of Yangliuqing.

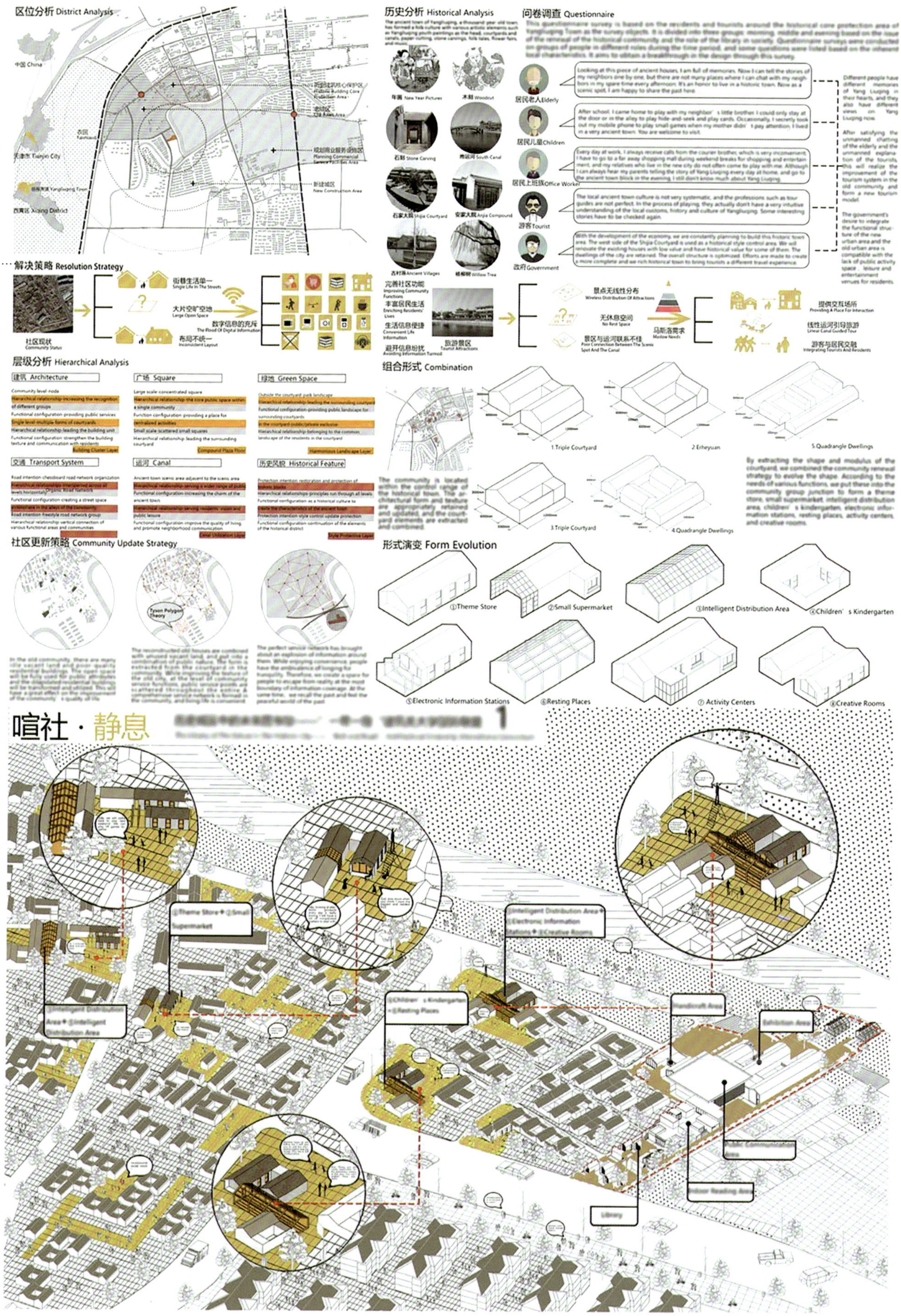
区位分析 District Analysis
历史分析 Historical Analysis
问卷调查 Questionnaire
解决策略 Resolution Strategy
层级分析 Hierarchical Analysis
组合形式 Combination
社区更新策略 Community Update Strategy
形式演变 Form Evolution
①Theme Store
②Small Supermarket
③Intelligent Distribution Area
④Children's Kindergarten
⑤Electronic Information Stations
⑥Resting Places
⑦Activity Centers
⑧Creative Rooms
喧社 · 静息

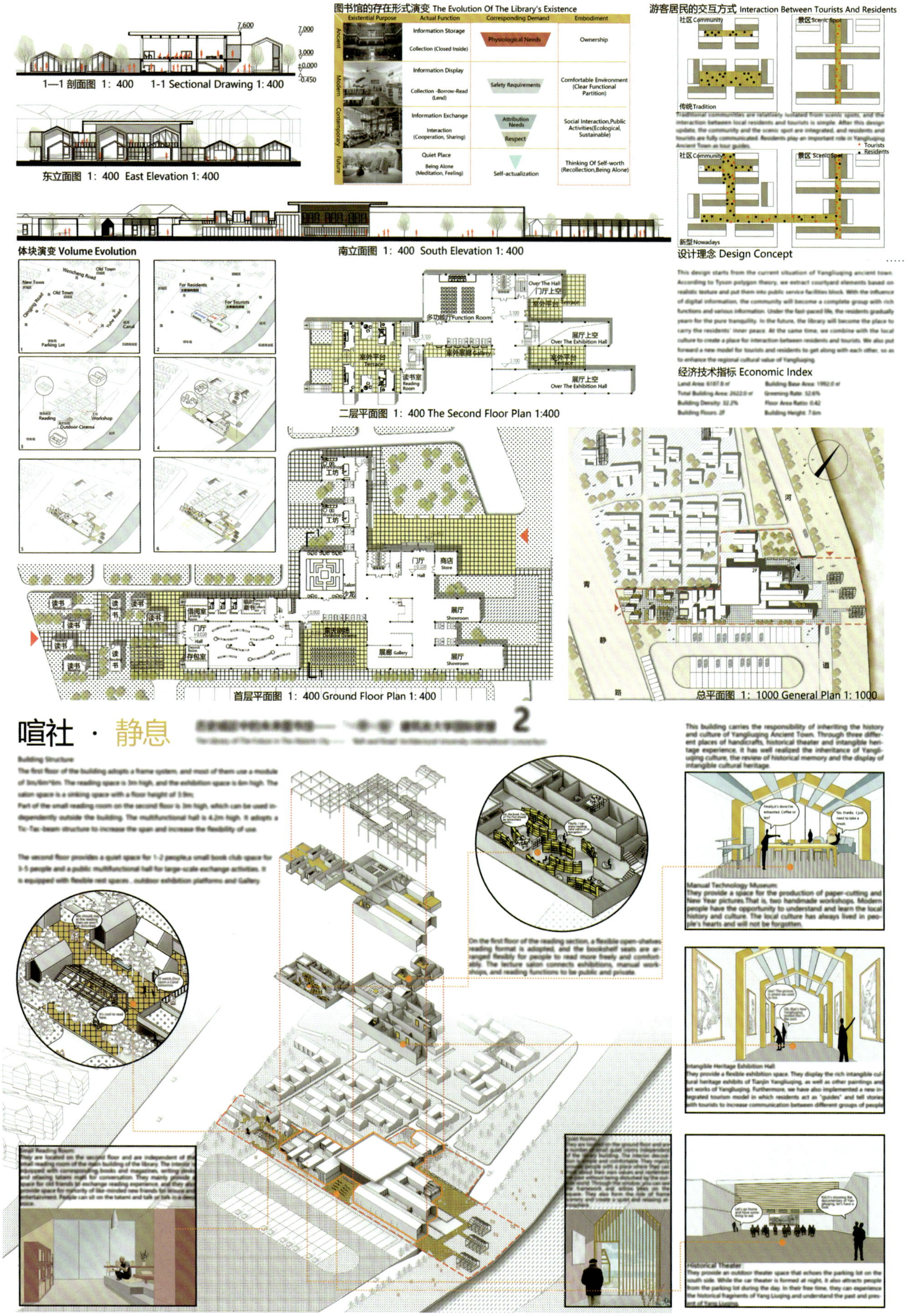
1—1 剖面图 1：400 1-1 Sectional Drawing 1: 400
7.600
7.000
3.000
±0.000
-0.450
东立面图 1：400 East Elevation 1: 400
图书馆的存在形式演变 The Evolution Of The Library's Existence
Existential Purpose
Actual Function
Corresponding Demand
Embodiment
Ancient
Modern
Contemporary
Future
Information Storage
Collection (Closed Inside)
Physiological Needs
Ownership
Information Display
Collection -Borrow-Read (Lend)
Safety Requirements
Comfortable Environment (Clear Functional Partition)
Information Exchange
Interaction (Cooperation, Sharing)
Attribution Needs
Respect
Social Interaction,Public Activities(Ecological, Sustainable)
Quiet Place
Being Alone (Meditation, Feeling)
Self-actualization
Thinking Of Self-worth (Recollection,Being Alone)
游客居民的交互方式 Interaction Between Tourists And Residents
社区 Community
景区 Scenic Spot
传统 Tradition
Tourists
Residents
新型 Nowadays
设计理念 Design Concept
体块演变 Volume Evolution
南立面图 1：400 South Elevation 1: 400
Old Town
Wenchang Road
New Town
Dingxing Road
Yuhe Road
Canal
Parking Lot
For Residents
For Tourists
Reading
Workshop
Outdoor Cinema
Over The Hall
门厅上空
室外平台 Terrace
多功能厅 Function Room
展厅上空 Over The Exhibition Hall
室外廊道 Gallery
读书室 Reading Room
二层平面图 1：400 The Second Floor Plan 1:400
经济技术指标 Economic Index
工坊 Workshop
门厅 Hall
商店 Store
沙龙 Salon
展厅 Showroom
展廊 Gallery
借阅室
存包室
读书
首层平面图 1：400 Ground Floor Plan 1: 400
河
青
静
路
道
总平面图 1：1000 General Plan 1: 1000
喧社 · 静息
2

■ 优秀奖 Honorable Mention

大隐于市

Hidden in the City

学校名称 University (College) Name
安徽建筑大学，中国
Anhui Jianzhu University, China

指导教师 Supervisor(s)
吴 敏 Wu Min 杨新刚 Yang Xin'gang

参赛学生 Participant(s)
白贤清 Bai Xianqing 丁卓群 Ding Zhuoqun
洪 玮 Hong Wei 徐 红 Xu Hong
张 郁 Zhang Yu

设计说明 Design Description

未来，图书馆的功能不再局限于书籍借阅与休闲，而是更注重通过多功能空间提高人的思想境界与道德素养。此作品将提供该社区所需的文化需求，与此同时，其独特的外观设计也将吸引居民与游人。项目选址在历史城区的居民重要聚集点——菜市场原址，拆除重建。菜市场平等、无障碍的沟通模式与图书馆的阅读空间相结合，体现“在生活与交流中寻找自我”的建筑理念。菜市场与图书馆的融合，是社会实践学习与理论知识学习的统一、自然与文明的碰撞、历史与未来的对话。

In the future, the function of the library is no longer limited to books dealing and leisure, but pay more attention to by multi-functional space to enhance people's ideology and moral accomplishment. This work will provide the cultural needs of the community, meanwhile, and its unique appearance will attract residents and visitors. Location in the historical city residents of the important point, the site, demolish rebuild. Combining the classless communication mode of the vegetable market with the reading space of the library, it embodies the architectural concept of "seeking self in life and communication". Vegetable market with the integration of library, it is the unity of the social practice and theory of knowledge learning, collision of nature and civilization, history and future of dialogue.

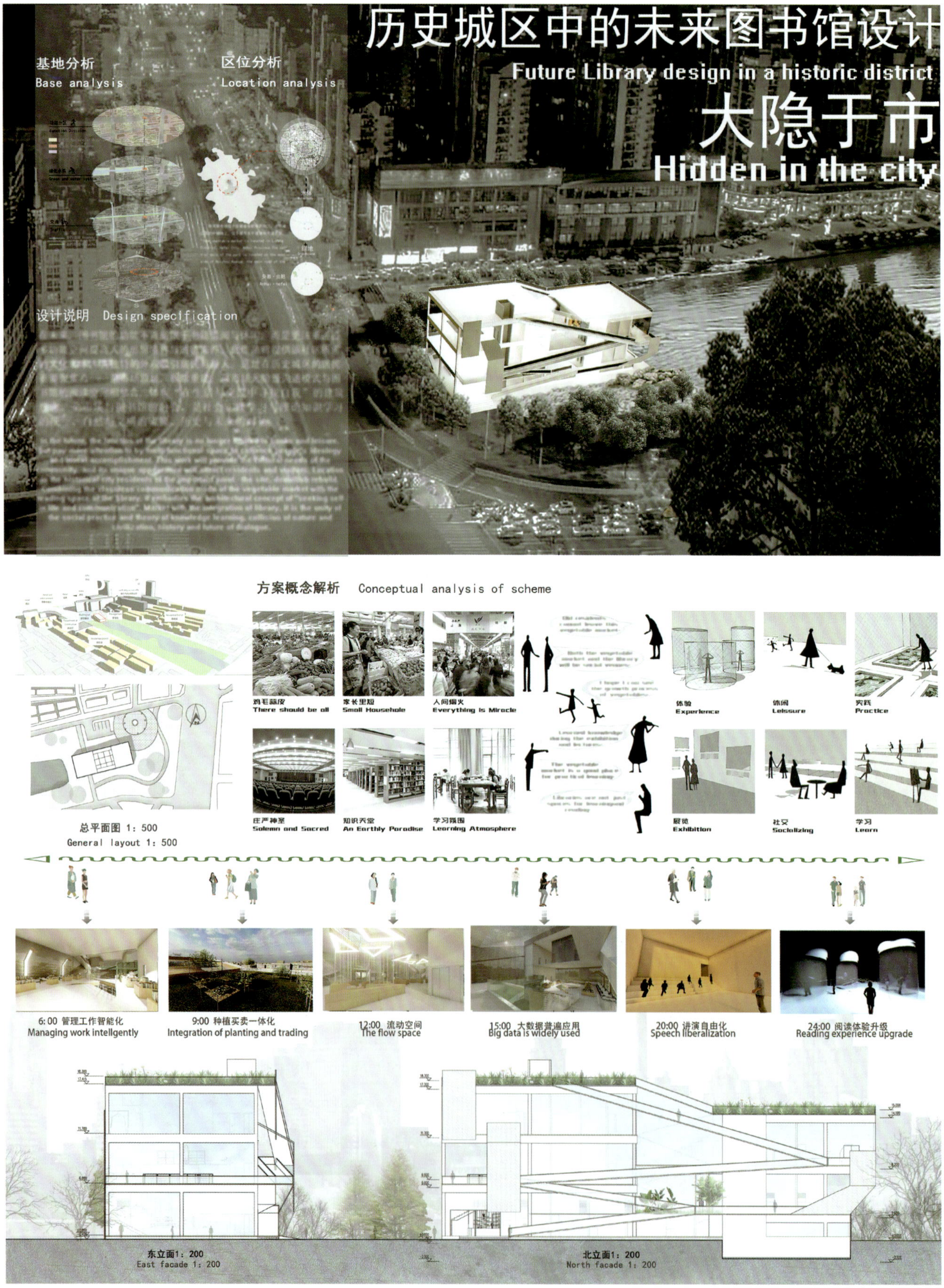
历史城区中的未来图书馆设计
Future Library design in a historic district
大隐于市
Hidden in the city
基地分析
Base analysis
区位分析
Location analysis
设计说明 Design specification
方案概念解析 Conceptual analysis of scheme
鸡毛蒜皮
There should be all
家长里短
Small Household
人间烟火
Everything is Miracle
庄严神圣
Solemn and Sacred
知识天堂
An Earthly Paradise
学习氛围
Learning Atmosphere
体验
Experience
休闲
Leisure
实践
Practice
展览
Exhibition
社交
Socializing
学习
Learn
总平面图 1：500
General layout 1：500
6:00 管理工作智能化
Managing work intelligently
9:00 种植买卖一体化
Integration of planting and trading
12:00 流动空间
The flow space
15:00 大数据普遍应用
Big data is widely used
20:00 讲演自由化
Speech liberalization
24:00 阅读体验升级
Reading experience upgrade
东立面1：200
East facade 1：200
北立面1：200
North facade 1：200

爆炸分析图
Explosion analysis

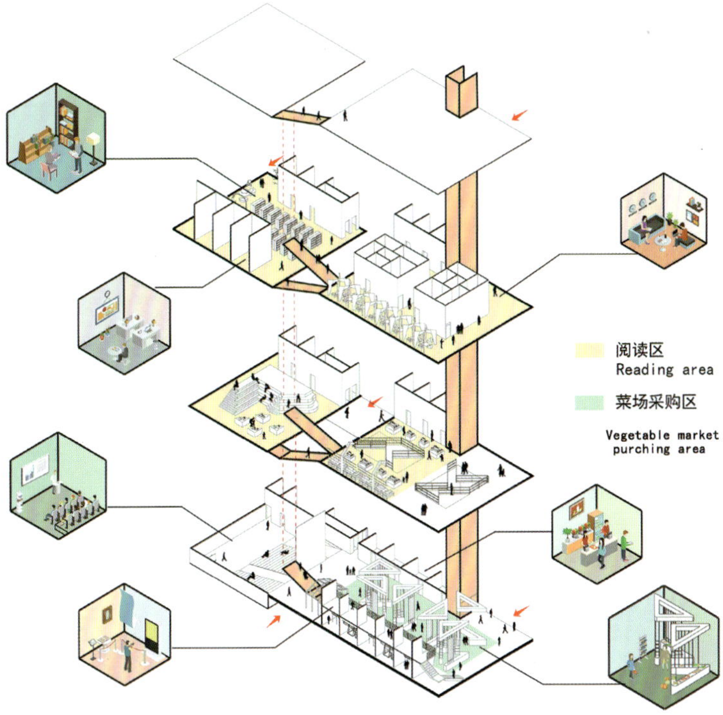

平面图1:350
The floor plan1: 350

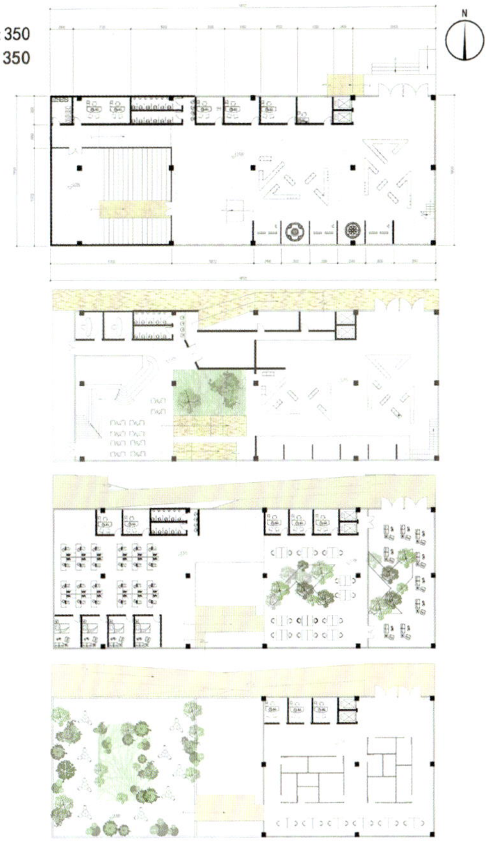

功能分析图 Function anaysis

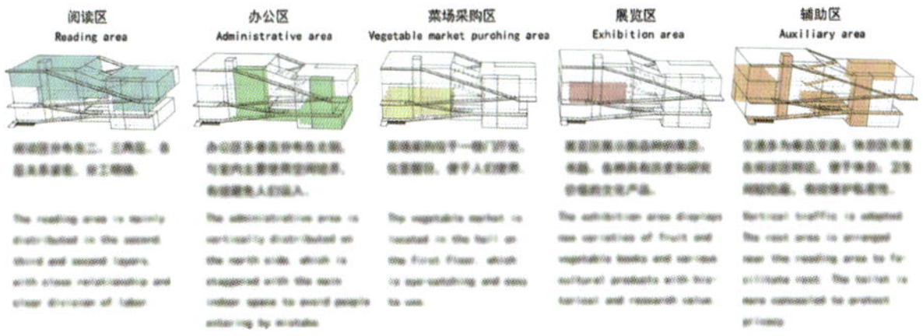

体块分析图 Block analysis

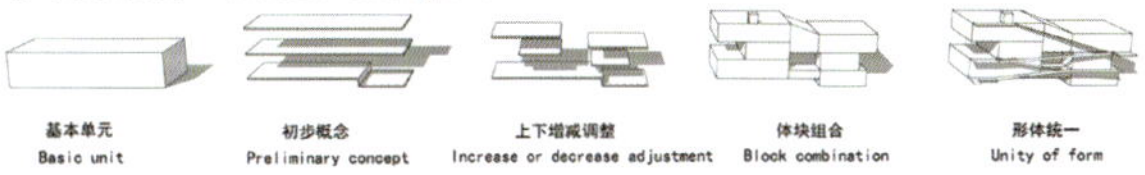

可持续发展分析图 Sustainable development

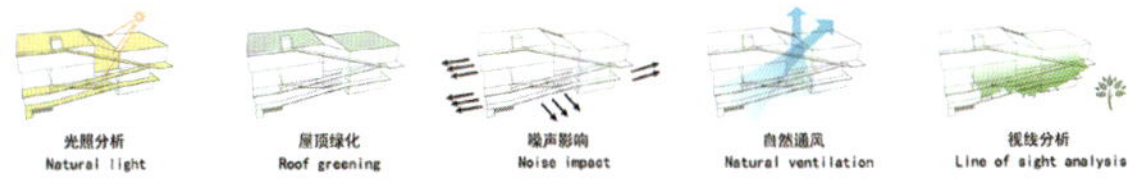

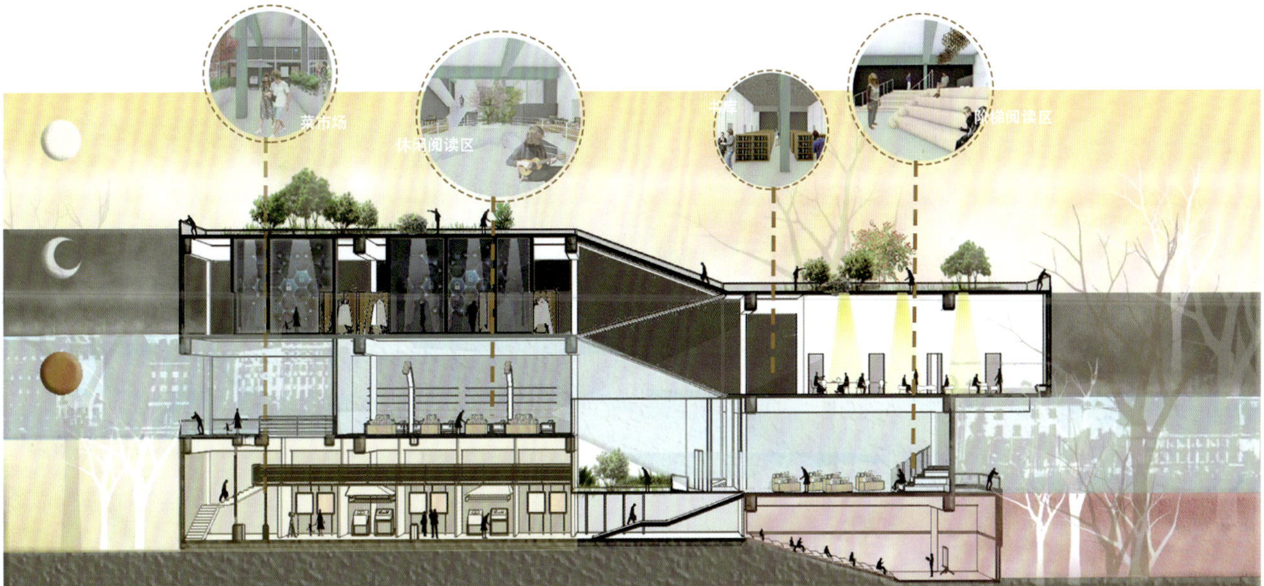

剖透视 Profile control perspective

■ 优秀奖 Honorable Mention

藏书桥廊——磁器口老城区未来图书馆设计

Book Collection Bridge Gallery — The Future Library Design of Ciqikou Old Town

学校名称 University (College) Name
重庆交通大学，中国
Chongqing Jiaotong University, China

指导教师 Supervisor(s)
温　泉 Wen Quan　　许　可 Xu Ke

参赛学生 Participant(s)
贾璐熙 Jia Luxi　　郭镇滔 Guo Zhentao
刘银露 Liu Yinlu　　刘人恺 Liu Renkai
石小满 Shi Xiaoman

设计说明 Design Description

场地选择在山水之城重庆的磁器口老城区。该地已有 1800 年的建镇历史，坐拥水陆交汇的地理环境，具有浓厚的历史气息。然而老城区建筑破旧，两岸交通不便，缺乏吸引居民驻足的公共空间，游离在邻里和江边的居民也渐渐互相疏远。事实甚至表明居住人口正在逐年流失。

基于串联两岸交通、激活公共空间的理念，我们将图书馆架设在江河支流上。作为桥建筑的图书馆，双向延伸了两岸的活动空间，将社区的居民与外来的游客凝聚在江面上。往来的人流在廊桥上相遇，徘徊的灵魂在书阁间驻足。

The site was chosen in the old town of Ciqikou, the city of mountains and rivers, Chongqing. This place has a history of 1800 years of being a town, it sits on a geographical environment where land and water meet and has a strong historical atmosphere. However, the old city has dilapidated buildings, inconvenient transportation on both sides of the strait, lack of public space to attract residents to stop, and residents living in the neighborhood and riverside are gradually alienating each other. Facts even show that the resident population is losing every year.

Therefore, out of the concept of connecting the two sides of the strait and activating public space, we set up the library on the tributaries of the river. As a bridge building, the library extends the activity space on both sides of the strait in both directions, condensing the residents of the community and the tourists from outside on the river. In the end, the flow of people meets on the bridge, and the wandering soul stops in the book pavilion.

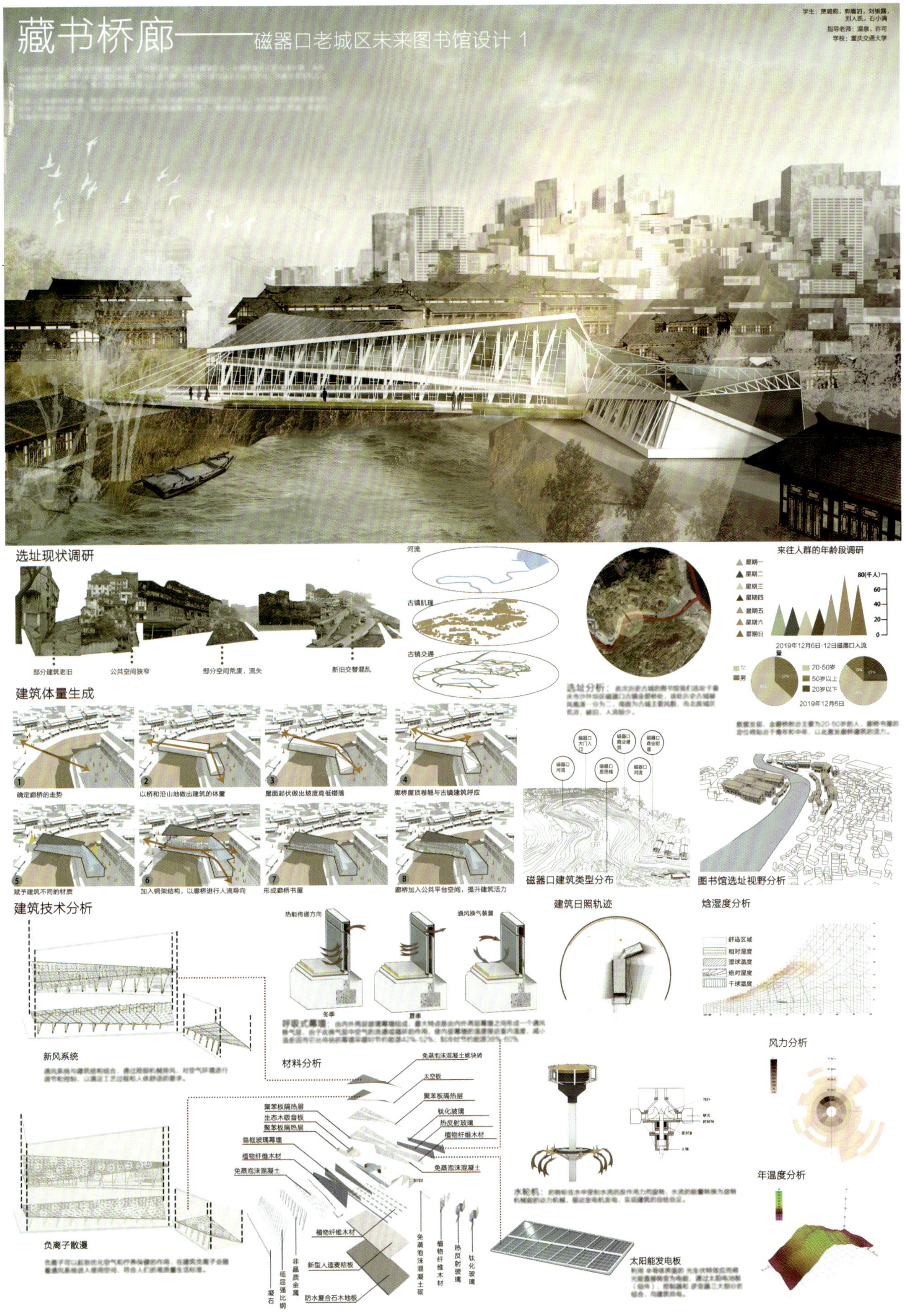
藏书桥廊——磁器口老城区未来图书馆设计 1
指导老师：温泉，许可
学校：重庆交通大学
选址现状调研
部分建筑老旧
公共空间狭窄
部分空间荒废，流失
新旧交替混乱
河流
古镇肌理
古镇交通
来往人群的年龄段调研
星期一
星期二
星期三
星期四
星期五
星期六
星期日
80(千人)
60
40
20
0
2019年12月6日-12日磁器口人流量
女
男
20-50岁
50岁以上
20岁以下
2019年12月6日
建筑体量生成
1 确定廊桥的走势
2 以桥和沿山地做出建筑的体量
3 屋面起伏做出坡度高低错落
4 廊桥屋顶卷翘与古镇建筑呼应
5 赋予建筑不同的材质
6 加入钢架结构，以廊桥进行人流导向
7 形成廊桥书屋
8 廊桥加入公共平台空间，提升建筑活力
磁器口建筑类型分布
图书馆选址视野分析
建筑技术分析
热能传递方向
通风换气装置
冬季
夏季
建筑日照轨迹
焓湿度分析
舒适区域
相对湿度
湿球温度
绝对湿度
干球温度
新风系统
材料分析
免蒸泡沫混凝土砌块砖
太空板
聚苯板隔热层
钛化玻璃
热反射玻璃
植物纤维木材
免蒸泡沫混凝土
聚苯板隔热层
生态木吸音板
聚苯板隔热层
隐框玻璃幕墙
植物纤维木材
免蒸泡沫混凝土
植物纤维木材
新型人造麦秸板
防水复合石木地板
凝石
低屈强比钢
非晶质金属
免蒸泡沫混凝土砌
植物纤维木材
热反射玻璃
钛化玻璃
负离子散漫
风力分析
年温度分析
太阳能发电板

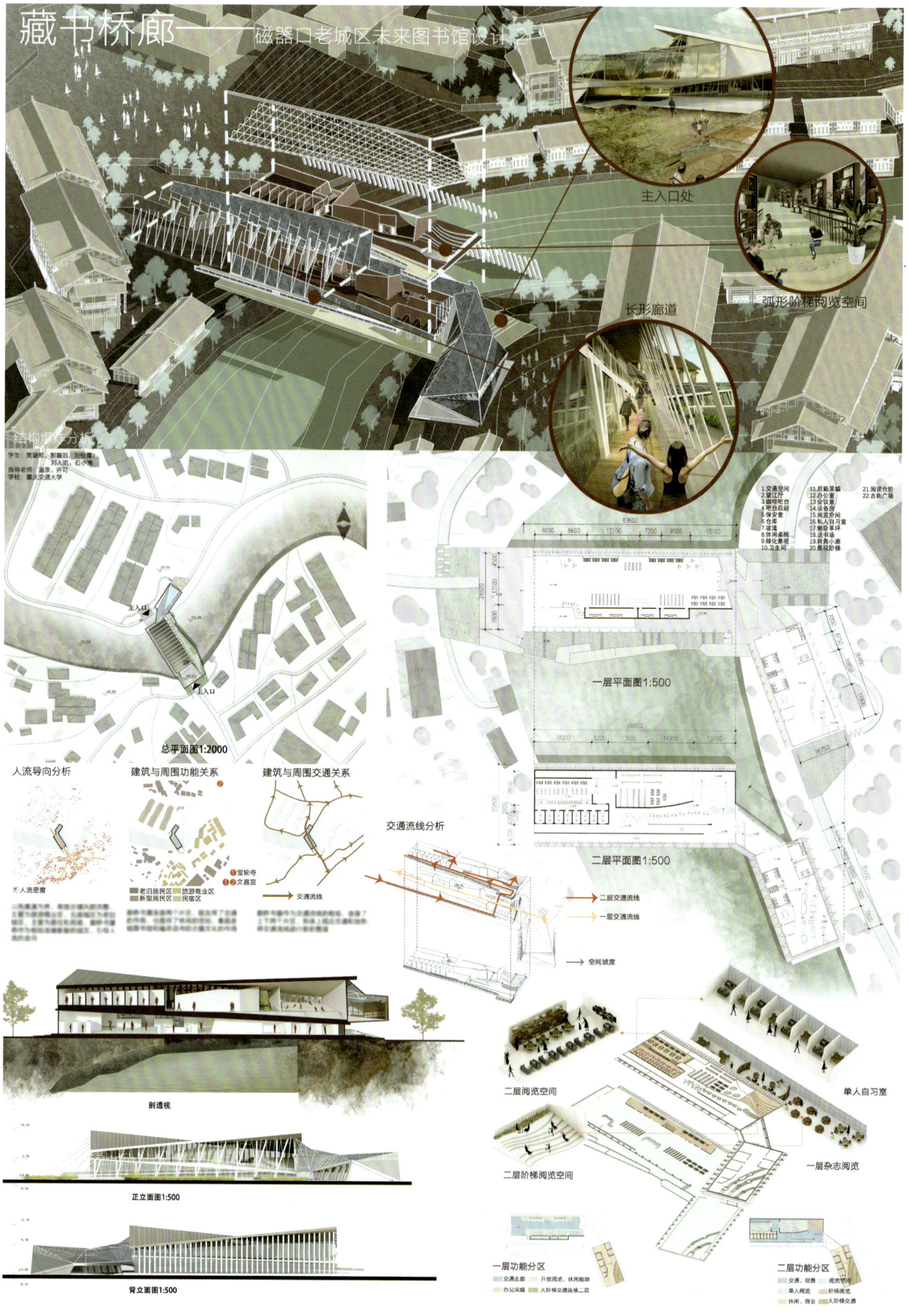
藏书桥廊——磁器口老城区未来图书馆设计 2
主入口处
弧形阶梯阅览空间
长形廊道
结构爆炸分析
指导老师：温泉，许可
学校：重庆交通大学
1.交通空间
2.望江厅
3.咖啡吧台
4.吧台后厨
5.保安室
6.仓库
7.坡道
8.休闲桌椅
9.绿化景观
10.卫生间
11.后勤采编
12.办公室
13.会议室
14.设备房
15.阅览空间
16.私人自习室
17.懒卧草坪
18.说书场
19.转角小廊
20.景观阶梯
21.阅读台阶
22.古街广场
主入口
总平面图1:2000
一层平面图1:500
二层平面图1:500
人流导向分析
人流密度
建筑与周围功能关系
宝轮寺
文昌宫
老旧居民区
旅游商业区
新型居民区
民宿区
建筑与周围交通关系
交通流线
交通流线分析
二层交通流线
一层交通流线
空间坡度
剖透视
正立面图1:500
背立面图1:500
二层阅览空间
单人自习室
二层阶梯阅览空间
一层杂志阅览
一层功能分区
二层功能分区

■ 优秀奖 Honorable Mention

"亚美尼亚圣拉扎罗岛"——图书馆改建方案 1

Saint Lazarus of the Armenians — Library Revalorisation Project 1

学校名称 University (College) Name
亚美尼亚国立建筑大学，亚美尼亚
National University of Architecture and Construction of Armenia, Armenia

指导教师 Supervisor(s)
Aloyan Artyom　　Sanamyan Hovhannes

参赛学生 Participant(s)
Vahram Aghajanyan　　Lernik Galstyan

设计说明 Design Description

首先，新建筑的外形复制了旧建筑，我们在屋顶、边角等重点位置上下功夫，立面和轮廓等整体设计上尽可能模仿威尼斯的旧建筑。

根据设计，图书馆部分占地约 500m^2，属于额外新增的建筑。其中 1 号和 2 号区域各占近一半面积。建筑一楼设咨询台和工作间，二楼设多个部分：公共档案室、历史手稿的特别储藏间、其他工作间。之所以把档案室设在二楼是出于防洪考虑，因为该地区经常发生洪灾。

关于餐厅部分，我们计划在邻近图书馆区域的上层设置玻璃餐厅，可以从楼梯直达，另外会有一个通向后厨的电梯。我们希望可以从餐厅所在的庭院看到大海，所以将餐厅设在更高、更开阔的上层，通往上层露台的通道即图书馆入口，这多少为新建筑增添了威尼斯风格。

由于目前的博物馆空间不足以容纳所有的展品，所以我们决定对其进行扩建。在目前内立面的基础上延伸屋角，由此获得更多空间。新的立面大部分选用玻璃材料，可以从建筑内部看到外面的风景，增强了建筑与环境的视觉联系。同时，我们使用走廊连接新旧庭院，方便游客自由移动。

现有的观景台区域状况并不良好，所以我们计划将其改造并作为博物馆部分的扩展，博物馆设楼梯、庭院设坡道通向这里。我们同样采用延伸屋角和玻璃立面的方法扩大了旧观景台空间，玻璃餐厅及露台空间可以自由支配，如果天气允许，完全可以在整理后用于露天展览。

First of all, we designed a new building, the plan shape of which is drawn by continuing the already existing (old building's) lines. In the facial part, we tried to keep the existing contours to the most possible extent, by designing the roof with the similar angles as in the old building, which will be obviously seen together in the view from Venice.

The library part, which according to the task, should be 500 m^2, is placed in this additional building. It is located on the 1st and 2nd, with approximately 300 m^2 area each. On the first floor the consultation area and the working rooms are placed. On the second floor there are several parts: the archive consultation room at public request, special reserves for historical manuscript, and additional working rooms. We located the archive rooms on the second floor, for safety purposes in case of flood, which is not seldom in this area.

For the cafeteria, we designed a glass box in the upper level, which has access via a few stairs from the library, and there is also an elevator, which is meant to serve for the kitchen. We wanted the inner yard, which is created by adding our new building, to have more than one access to the sea, so, by taking the cafeteria box to the higher level and making openings we made it easier to get to the sea, and also used the top part the passage as a terrace with entrance from the library. This also provided some Venice style to the new constructions.

Then we decided to expand the museum part, because the area of the current space is not large enough for all the exponents. So, we kept the existing facade in the interior part and added a new one by continuing the angle of the old building's roof. The new facade is mostly glass made, which helps to create some visual connection of the building to the environment, with beautiful views opened from the inside. Also, we thought that it would be good to connect the existing inner yard with ours, for making the movement of the visitors easier.

There is an existing terrace near the observatory which

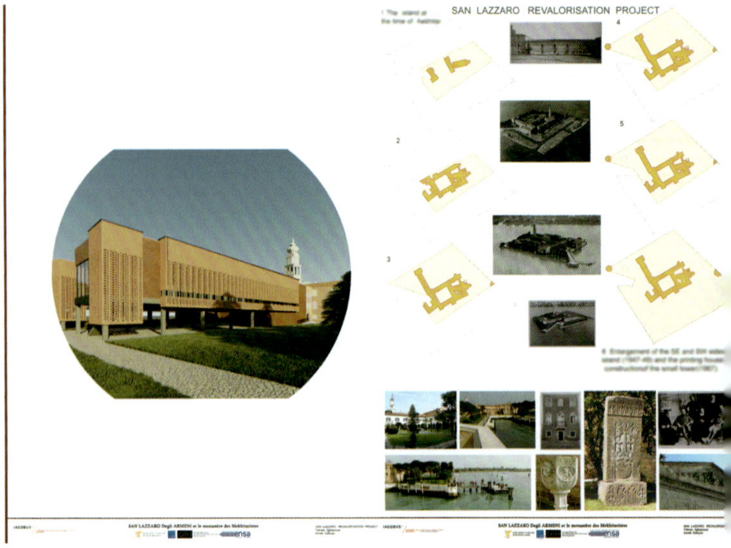

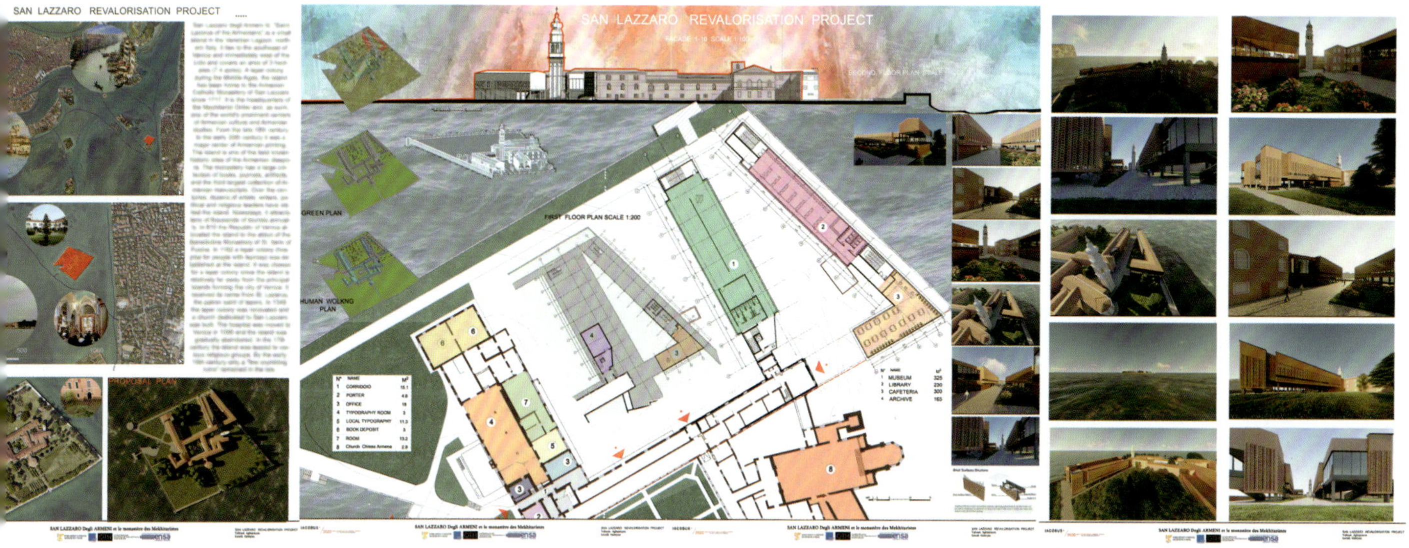

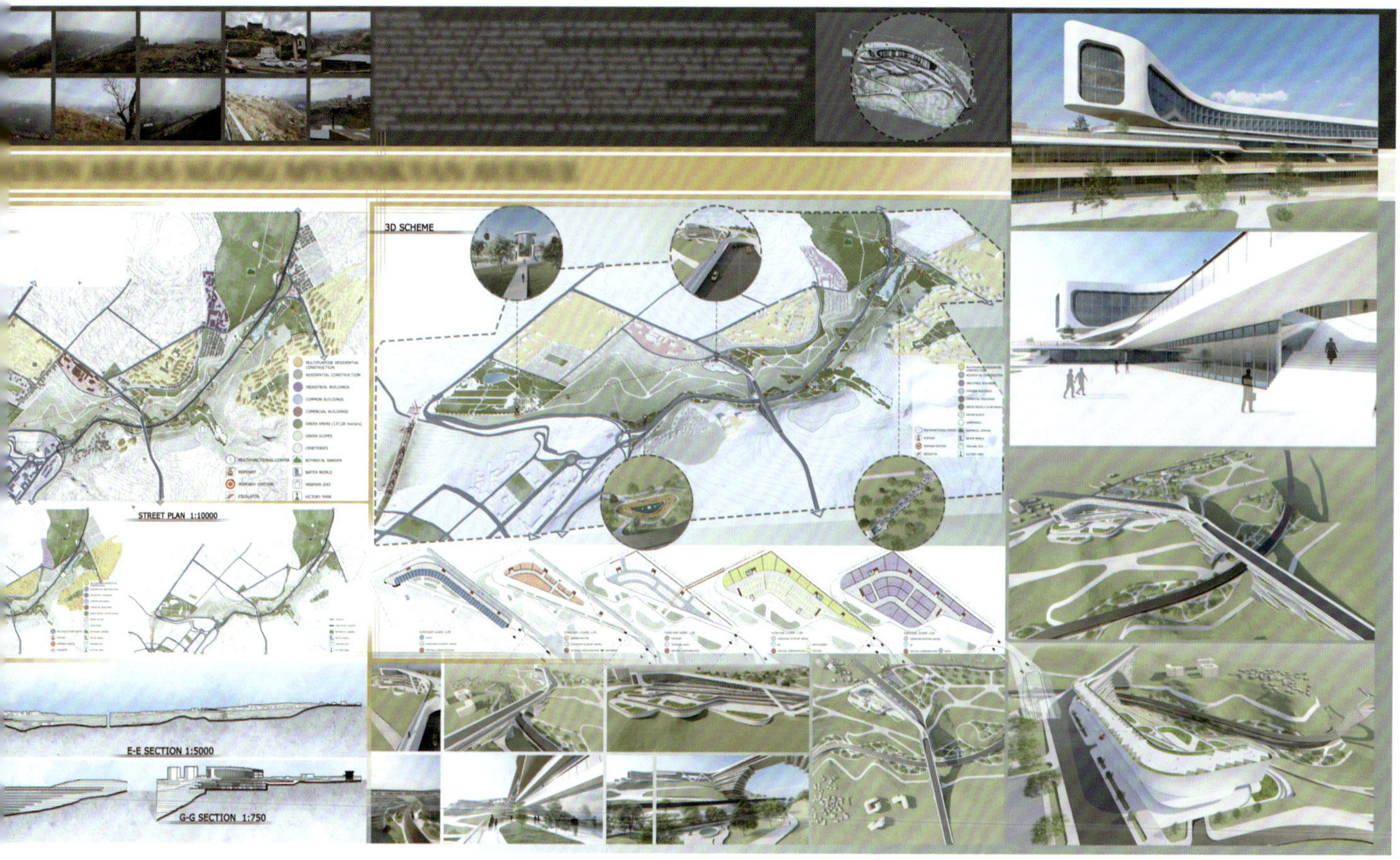

is not in a good condition so we decided to reconstruct it and to use it as a continuation of the museum part. Here we also keep the angle of the existing roof continuing it with glass panels. The space between this glass box and the open terrace is highly flexible and easy to configure and reconfigure for open air exhibition when the weather allows.

■ 优秀奖 Honorable Mention

梁赞市图书馆

The Library Project in the City of Ryazan

学校名称 University (College) Name
莫斯科理工大学梁赞分校，俄罗斯
Ryazan Institute (branch) Moscow Polytechnic University, Russia

指导教师 Supervisor(s)
Koghnova Anna Alekseevna

参赛学生 Participant(s)
Seliverstova Elizaveta

设计说明 Design Description

该图书馆方案旨在打造现代化的文娱场所。

方案展现了图书馆这一公共场所的未来发展潜力。尽管书籍作为纸质印刷物已经不再辉煌，但人们对新知识的渴望和在阅读中自学的方式仍存。

图书馆将变成字面意义上的公共场所，而我们希望借此机会，在这历史悠久的城市中心，实践将生态融于建筑的理念，打造开放、环保、和谐、高科技、符合人体工程学的新型图书馆。本方案反映了 20 世纪中期以来的建筑发展趋势，即跟随城市发展，让建筑不断“完整”。

未来的图书馆要体现人与自然统一的理念，所以我们的方案中设有开放的露台，可以俯瞰宽阔的水面，有科学证据表明，宽广的空间有利于人消化信息，我们的设计理念正是要突出这一点。图书馆内部强调交互性，要安装最先进的 IT 设备，确保空气流通，营造适合信息工作和文化活动的氛围。

我们设计的新型图书馆别出心裁，打造舒适和放松的环境，帮助人们消化信息。长此以往，不仅有助于居民保持心理健康，还能提高人口文化素质。

The concept of the library design is a modern space for learning and cultural leisure.

This concept reveals the library as a public space of the future with unlimited possibilities. It is worth noting that reading, self-education and a craving for new knowledge do not lose relevance, despite the fact that the book, as a

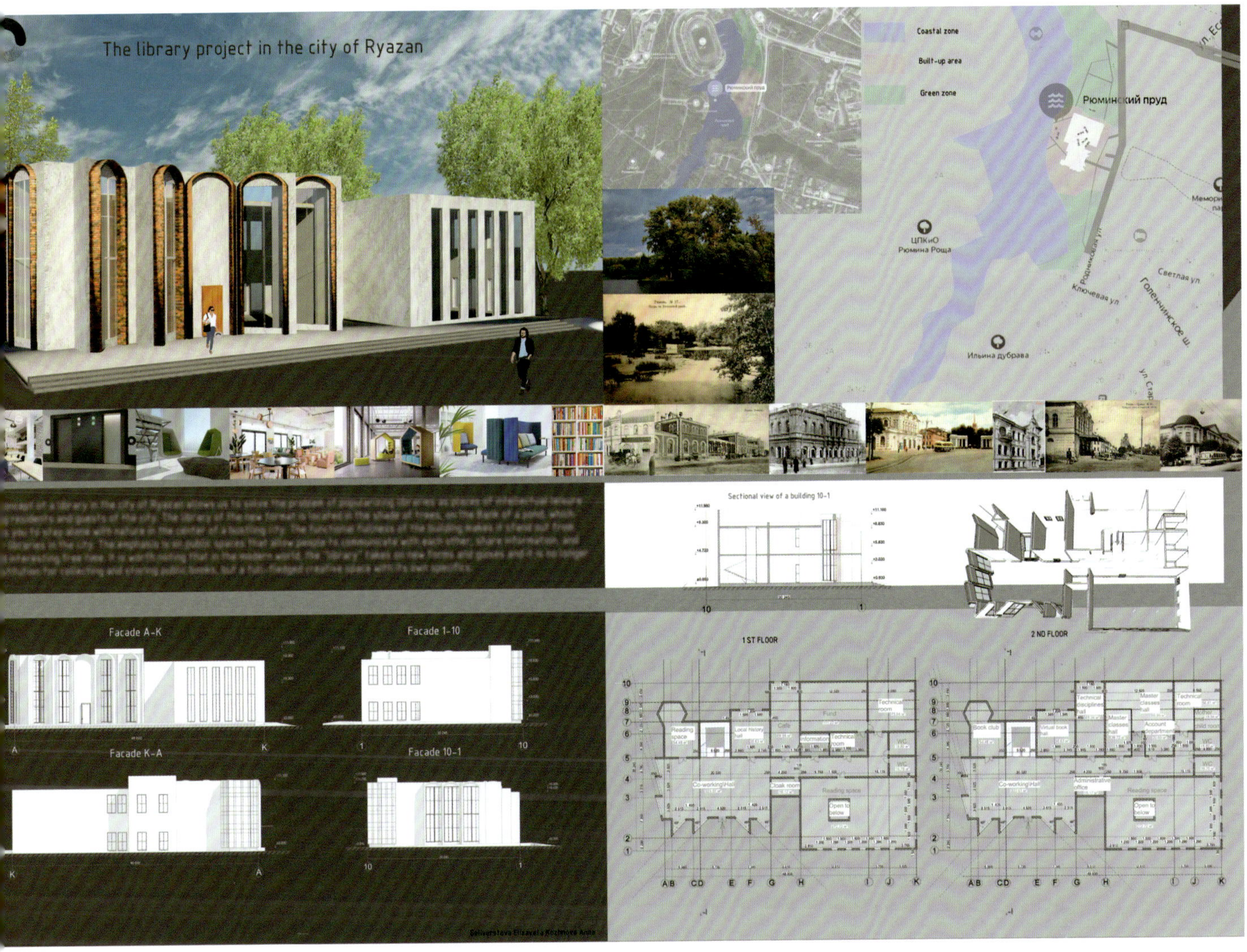

printed publication, is gradually losing its popularity.

Thus, the library becomes precisely public space in the broad sense of the word. The idea to fit architecture into the context of wildlife in the center of the historical city dictates some conditions: openness, environmental friendliness, harmony, technicality and ergonomics. This project reflects the architectural trends of the mid-20th century and allows you to leave the architectural form "incomplete" for further dialogue with the changing architecture of the city.

The concept of the library of the future with open terraces overlooking the expanse of the water surface carries the idea of unity with nature, which from a scientific point of view favorably affects the digestibility of information, and in the concept of library space this is an important factor. The interior part of the project implies interactivity, equipment with the latest technologies in the field of IT, airiness and a cozy atmosphere for working with information or for cultural activities.

Thus, the concept of this design is interesting from the point of view of presenting a new format of library space, which allows a person to feel comfortable and relaxed in the environment of receiving new information, which will undoubtedly have a beneficial effect on the psycho-emotional state of a person and help to increase the level of cultural quality of the population.

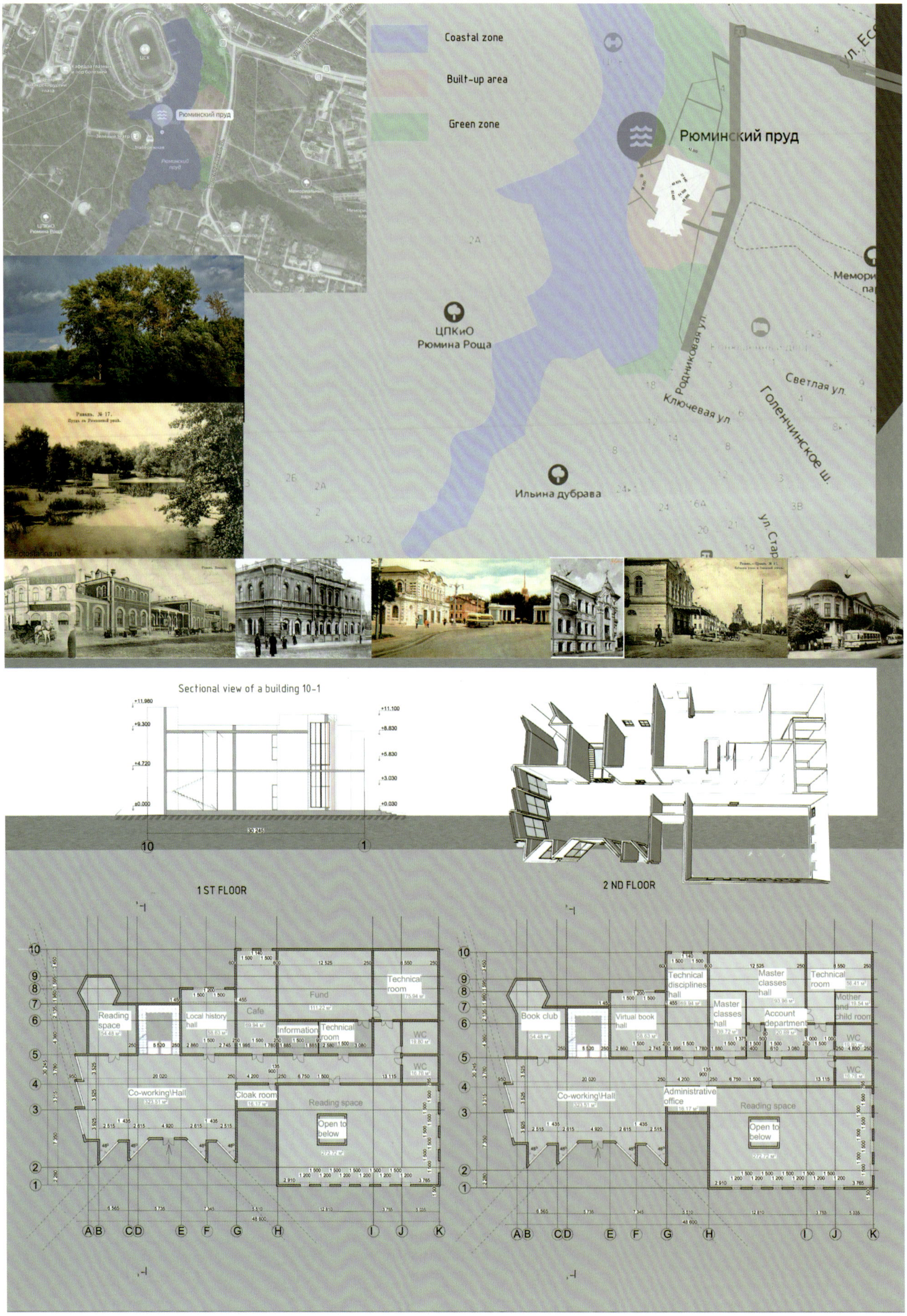
Coastal zone
Built-up area
Green zone
Рюминский пруд
ЦПКиО
Рюмина Роща
Ильина дубрава
Родниковая ул.
Ключевая ул
Светлая ул.
Голенчинское ш.
Sectional view of a building 10-1
1 ST FLOOR
2 ND FLOOR
Reading space
Local history hall
Cafe
Fund
Technical room
Information
WC
Co-working\Hall
Cloak room
Open to below
Book club
Virtual book hall
Technical disciplines hall
Master classes hall
Account department
Mother and child room
Administrative office

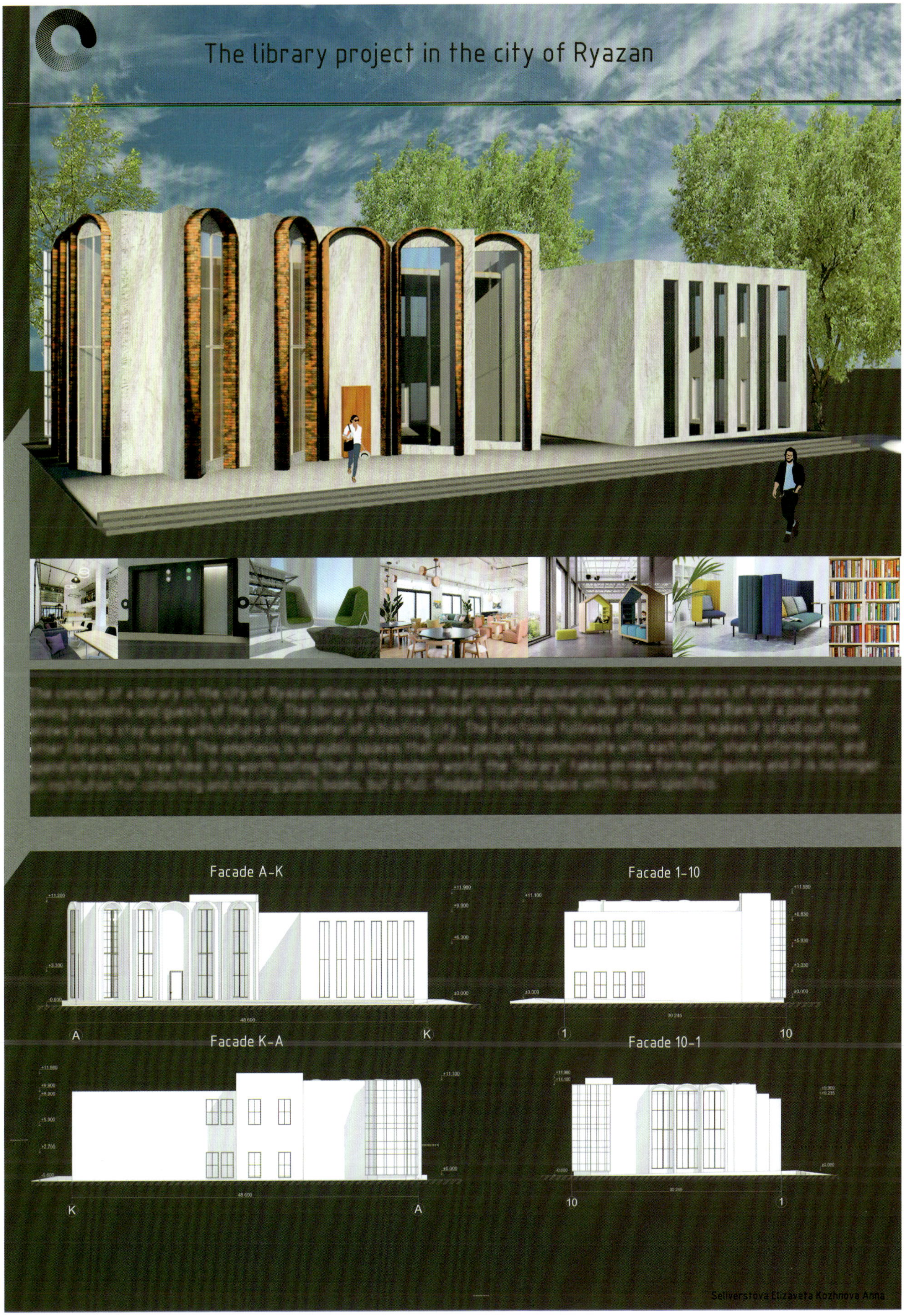
The library project in the city of Ryazan
Facade A-K
Facade 1-10
Facade K-A
Facade 10-1
Seliverstova Elizaveta Kozhnova Anna

优秀奖 Honorable Mention

华沙普拉斯基港图书馆

City Library at the Port Praski in Warsaw

学校名称 University (College) Name
华沙生态与管理大学，波兰
University of Ecology and Management in Warsaw, Poland

指导教师 Supervisor(s)
Agnieszka Starzyk

参赛学生 Participant(s)
Gabriela Anna Żarnowska

设计说明 Design Description

数字化技术的发展使图书馆的生存环境日渐严峻。对实体书需求的减少使图书馆失去立足之本，新技术带来新挑战，对图书馆提出新要求。要拯救图书馆，我们就必须行动起来，帮助图书馆跟上时代的脚步。我们要打破图书馆只能用于藏书的刻板印象，强调其作为服务机构的职能。图书馆需要满足读者的需求，而读者需要的远不只一本书那么简单。读者真正需要的是现代化的资源搜索服务和文化机构所能提供的会议和文娱场所。

虽然本设计方案采用极简的建筑形式，但其空间布局上大有乾坤，不会限制读者的行动。方案以新型餐厅、休闲区、会议室和机房为蓝图，勾勒出未来的图书馆场景。建筑布局经过精心设计，为阅读提供了安静和集中的环境，同时规划了娱乐、休闲、社交的专门区域。未来图书馆不再需要排队借书，先进的识别技术与智能手机可以让这一切变得轻松。然而，图书馆的实际情况与理想截然相反，目前它仍是一个功能单一的建筑，内部拥挤，积灰落土。

普拉斯基港的历史可以追溯到 1901 年。时至今日，普拉斯基港已然改头换面，但建筑设计与环保要求仍难两全。因此，我们计划使用绿色建材，低负担的建筑形式和环保墙体将代表城市与自然的和谐，诠释建筑是如何与科技携手并进的。

Along with the development of digitalization, a question about the sense of the existence of libraries arose. Since the need of physical existence of books was undermined so was the existence of library edifices. Undoubtedly the approaching technological breakthrough brings even newer challenge and forces people to take action which let them implement solutions that will prevent getting rid of libraries and will allow them to develop with time. It should be assumed that the library is not only a house of books but first of all an institution that provides services. Its activity is closely connected with the readers' needs which do not end with the need to get a sought book. The reader looks for modern methods of accessing resources, as well as a place of meetings and entertainment which is given by the culture sector.

This work presents a minimalistic architectural form which is a consequence of well thought-over space, where the user can move around freely. This design shows the introduction of new functions in the form of restaurant and relaxation spaces, conference rooms as well as appointed computer area which constitute an image of a future library. The well-thought layout of the building provides a calm, concentrated environment for reading books and also an appointed area for entertainment, shared relax and possibility of communication. Furthermore, the library of the future does not require from the readers queuing to the lending room, with making use of modern scan methods, it is enough to use a smartphone. All this is in opposition to the image of the library as a non-functional building consisting of cramped, dusty interiors.

Port Praski dates back in history to 1901. The numerous changes of the port led to its closure and appearing of a new environment which arises arguments among building designers and environmentalists. Using natural materials, a light form of the building and green walls create a balanced ecosystem which fits in the urban landscape and proves how architecture goes hand in hand with the technology development.

CITY LIBRARY AT THE PORT PRASKI IN WARSAW

I/II

THE ROAD TRAFFIC ANALYSIS

- AREA
- BUS ROUTES
- METRO ROUTES
- RAILWAY ROUTES
- BUS STATION
- METRO STATION
- RAILWAY STATION

VISIBILITY ANALYSIS

VEGETATION ANALYSIS

PEDESTRIAN TRAFFIC ANALYSIS

BIKE TRAFFIC ANALYSIS

GROUND FLOOR PLAN SCALE 1:200

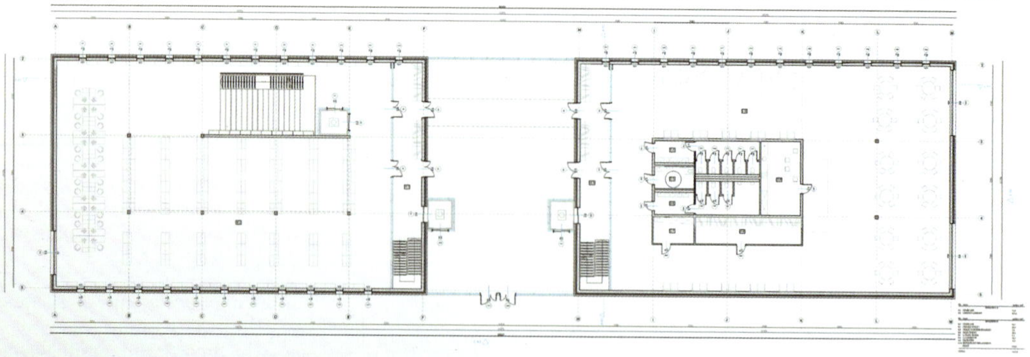

BASEMNET FLOOR PLAN SCALE 1:400

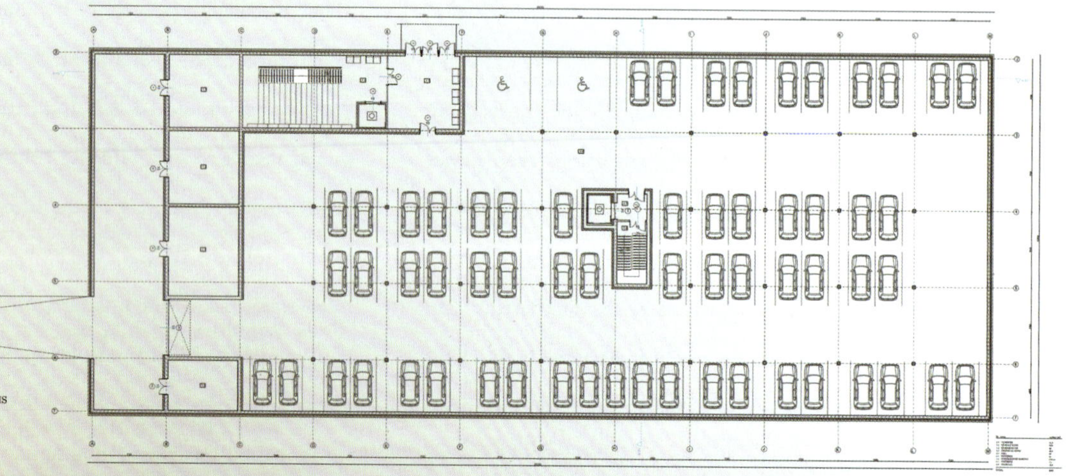

SECTION A-02 SCALE 1:200

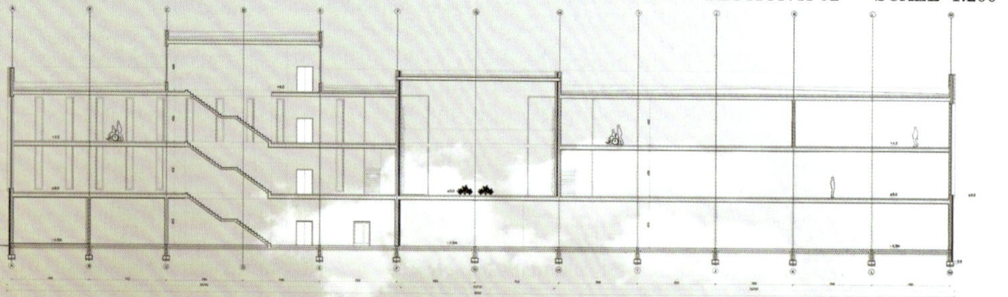

CITY LIBRARY AT THE PORT PRASKI IN WARSAW

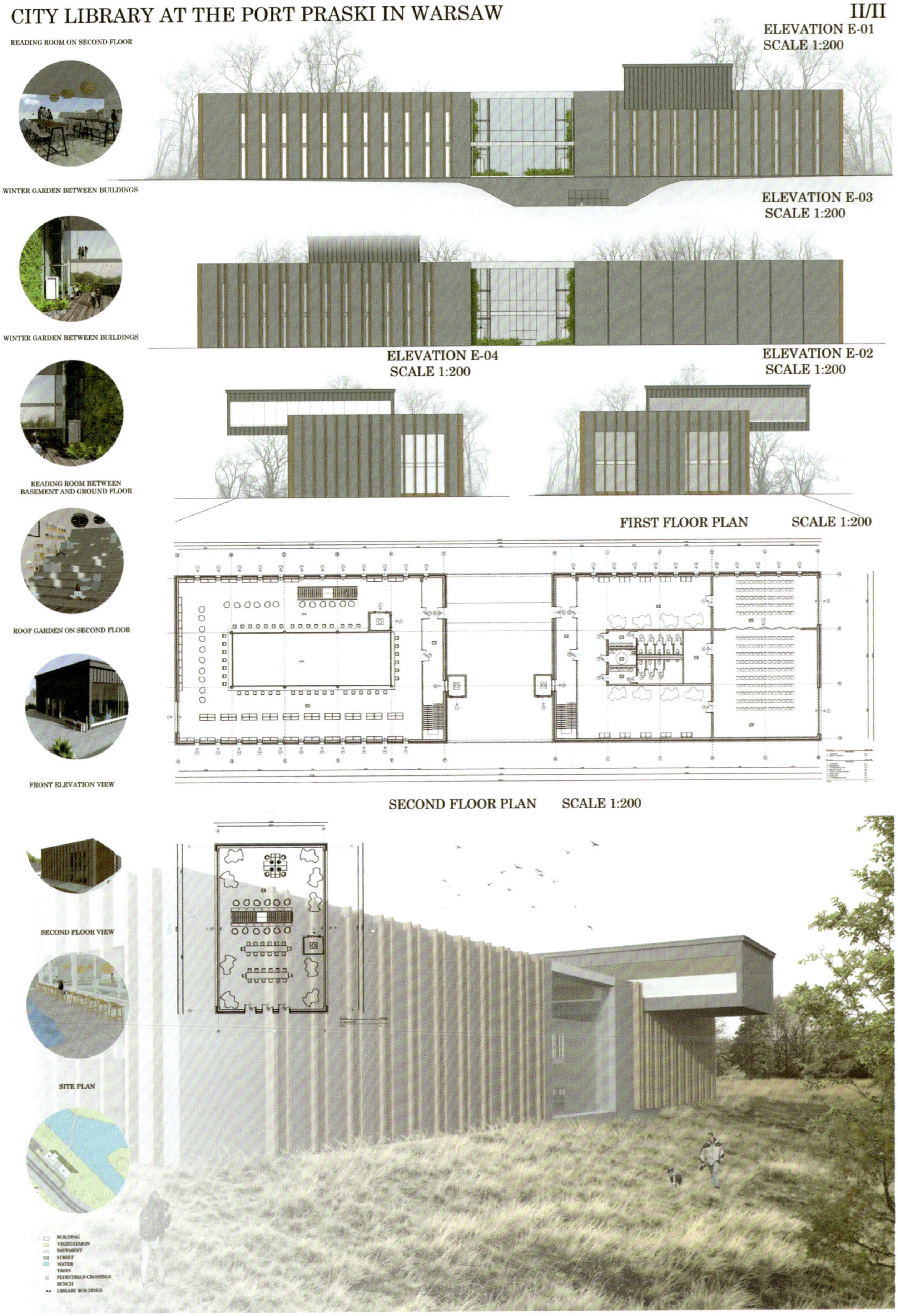

■ 其他作品　Other Works

记忆容器

Memory Container

学校名称　University (College) Name
长春建筑学院，中国
Changchun University of Architecture and Civil Engineering, China

指导教师　Supervisor(s)
李　冲　Li Chong　　李建英　Li Jianying

参赛学生　Participant(s)
白贤清　Bai Xianqing　　丁卓群　Ding Zhuoqun
洪　玮　Hong Wei　　徐　红　Xu Hong
张　郁　Zhang Yu

设计说明 Design Description

随着时代的发展与科技的进步，城市的环境被治理得越来越好，社会生产力也越来越发达，人们的生活水平大大提升。为适应社会的发展，大量承载历史记忆的老旧建筑被拆除，逐步走向“千城一面”，人们出现了记忆缺失的现象。我们希望未来图书馆不仅在功能、经济、生态方面有所改进，同时起到唤醒人们内心深处记忆的作用，给予人们精神上的力量。

With the development of the times and the progress of science and technology, the urban environment has been governed better and better, social productive forces have become more and more developed, and people's living standards have been greatly improved. In order to adapt to the development of the society, a large number of old buildings bearing historical memory have been demolished, making more similarities and deficient historical memory of cities. Therefore, we hope that the future library will not only improve in function, economy and ecology, but also play a role in awakening people's deep memory and giving people spiritual strength.

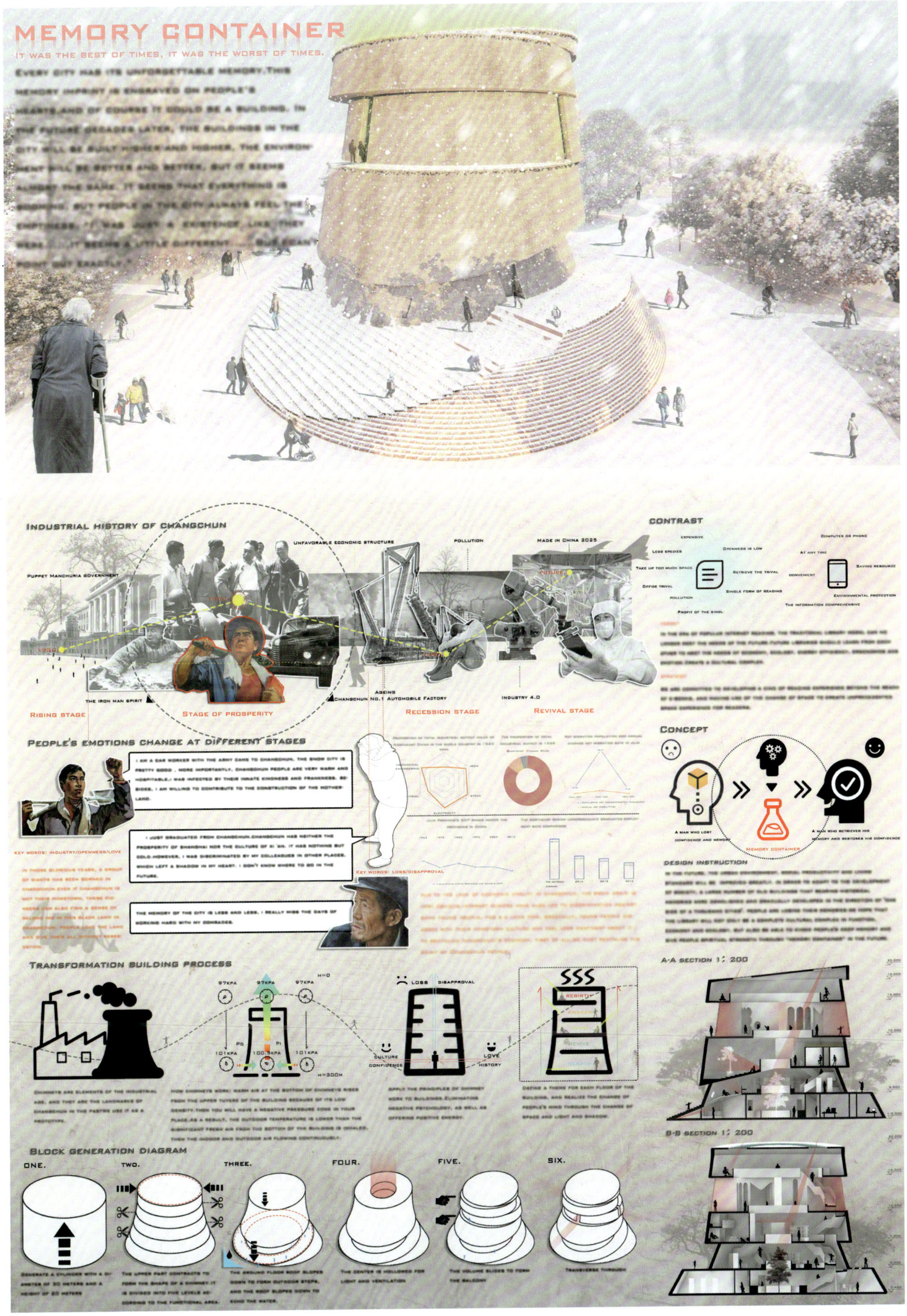
MEMORY CONTAINER
IT WAS THE BEST OF TIMES, IT WAS THE WORST OF TIMES.
INDUSTRIAL HISTORY OF CHANGCHUN
PUPPET MANCHURIA GOVERNMENT
UNFAVORABLE ECONOMIC STRUCTURE
POLLUTION
MADE IN CHINA 2025
FUTURE
THE IRON MAN SPIRIT
AGEING
CHANGCHUN NO.1 AUTOMOBILE FACTORY
INDUSTRY 4.0
RISING STAGE
STAGE OF PROSPERITY
RECESSION STAGE
REVIVAL STAGE
CONTRAST
CONCEPT
A MAN WHO LOST CONFIDENCE AND MEMORY
MEMORY CONTAINER
A MAN WHO RETRIEVES HIS MEMORY AND RESTORES HIS CONFIDENCE
DESIGN INSTRUCTION
PEOPLE'S EMOTIONS CHANGE AT DIFFERENT STAGES
KEY WORDS: INDUSTRY/OPENNESS/LOVE
KEY WORDS: LOSS/DISAPPROVAL
TRANSFORMATION BUILDING PROCESS
97KPA
101KPA
100.5KPA
H=0
H=300M
LOSS
DISAPPROVAL
CULTURE
CONFIDENCE
LOVE
HISTORY
REBIRTH
REVIVE
A-A SECTION 1: 200
B-B SECTION 1: 200
BLOCK GENERATION DIAGRAM
ONE.
TWO.
THREE.
FOUR.
FIVE.
SIX.

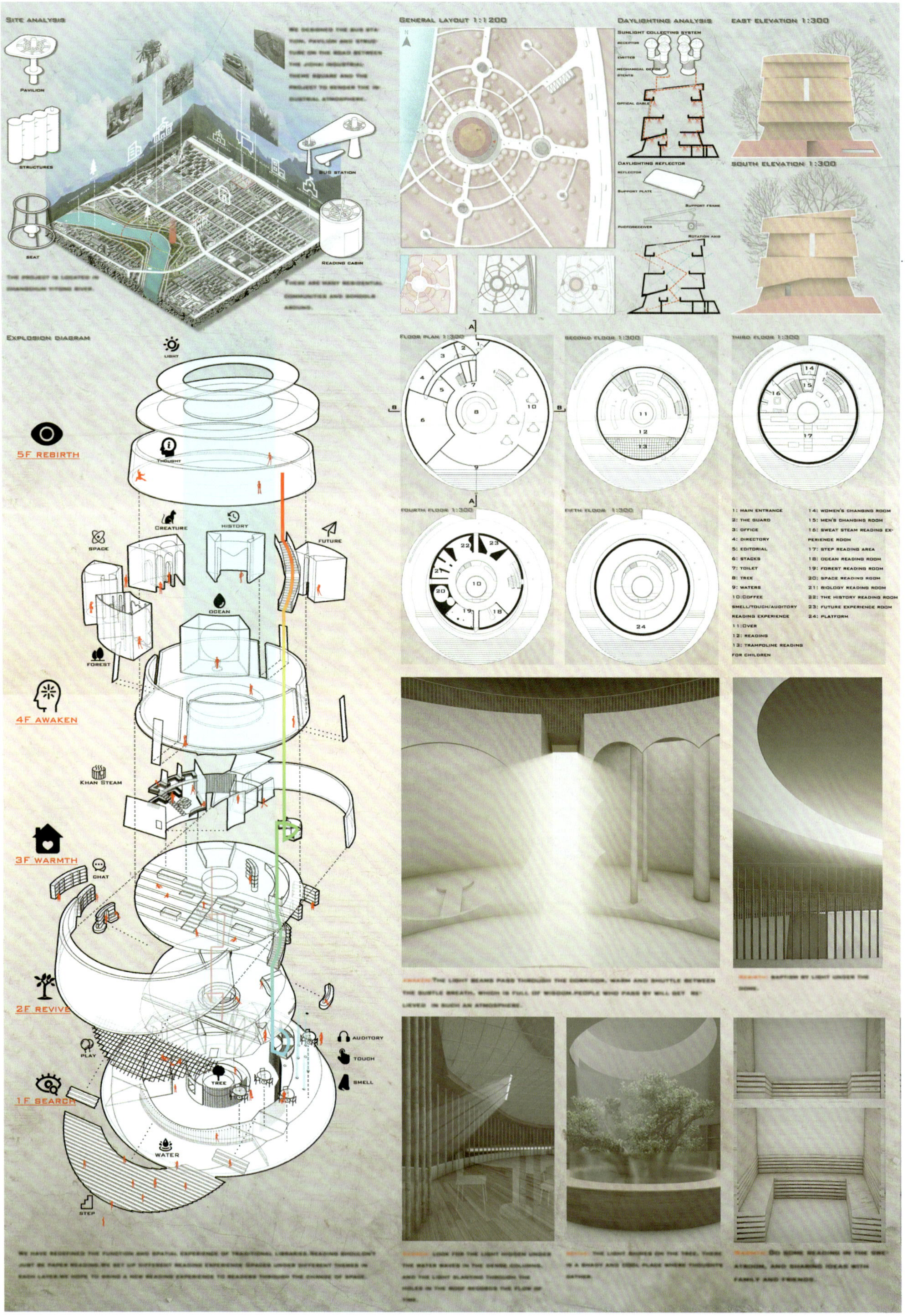
SITE ANALYSIS
PAVILION
STRUCTURES
SEAT
BUS STATION
READING CABIN
GENERAL LAYOUT 1:1200
DAYLIGHTING ANALYSIS
SUNLIGHT COLLECTING SYSTEM
OPTICAL CABLE
DAYLIGHTING REFLECTOR
REFLECTOR
SUPPORT PLATE
SUPPORT FRAME
PHOTORECEIVER
ROTATION AXIS
EAST ELEVATION 1:300
SOUTH ELEVATION 1:300
EXPLOSION DIAGRAM
LIGHT
THOUGHT
5F REBIRTH
CREATURE
HISTORY
SPACE
FUTURE
OCEAN
FOREST
4F AWAKEN
KHAN STEAM
3F WARMTH
CHAT
2F REVIVE
PLAY
TREE
AUDITORY
TOUCH
SMELL
1F SEARCH
WATER
STEP
FLOOR PLAN 1:300
SECOND FLOOR 1:300
THIRD FLOOR 1:300
FOURTH FLOOR 1:300
FIFTH FLOOR 1:300
1: MAIN ENTRANCE
2: THE GUARD
3: OFFICE
4: DIRECTORY
5: EDITORIAL
6: STACKS
7: TOILET
8: TREE
9: WATERS
10:COFFEE
SMELL/TOUCH/AUDITORY
READING EXPERIENCE
11:OVER
12: READING
13: TRAMPOLINE READING
FOR CHILDREN
14: WOMEN'S CHANGING ROOM
15: MEN'S CHANGING ROOM
16: SWEAT STEAM READING EXPERIENCE ROOM
17: STEP READING AREA
18: OCEAN READING ROOM
19: FOREST READING ROOM
20: SPACE READING ROOM
21: BIOLOGY READING ROOM
22: THE HISTORY READING ROOM
23: FUTURE EXPERIENCE ROOM
24: PLATFORM

■ 其他作品 Other Works

予福图书馆

Blessing Library

学校名称 University (College) Name
河南城建学院，中国
Henan University of Urban Construction, China

指导教师 Supervisor(s)
樊子铭 Fan Ziming　　王晓宁 Wang Xiaoning

参赛学生 Participant(s)
朱明坤 Zhu Mingkun　　姚雪峰 Yao Xuefeng
李　慧 Li Hui　　禹　琪 Yu Qi

设计说明 Design Description

本设计旨在创建一个与活动场地有机结合的未来图书馆。通过图书馆的设立，使活动场地和周围环境更加具有活力。通过运用一些新的技术和创新理念，打造一个具有新功能形式的未来图书馆。未来图书馆既可以解决现有图书馆的问题，又可以赋予图书馆新的活动形式和开设方式，通过新的设计方法去表达图书馆所包含的文化价值和图书馆的美学形式。

The design aims to create a future library organically integrated with the venue. The establishment of the library will invigorate the venue and its surroundings. Based on some new technologies and innovative ideas, we can build a future library characterized by new functions and forms. The future library can not only solve the problems of the existing library, but also endow the library with new forms of activity and a new method of establishment, to express the inner culture values and aesthetics form of library through new method of design.

予福 图书馆建筑设计 Blessing
首层平面图 1：250
主入口
次入口
设计说明：

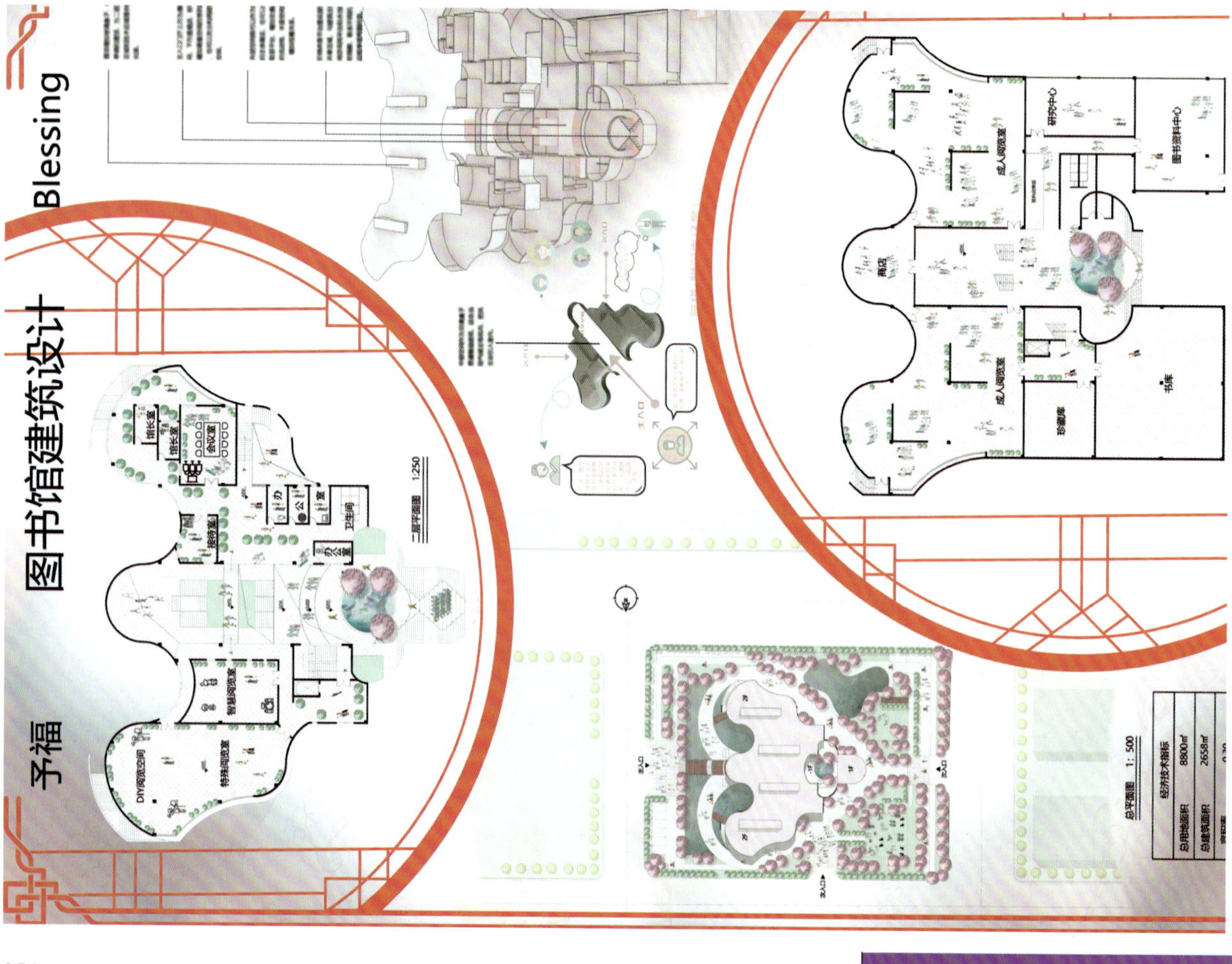
予福 图书馆建筑设计 Blessing
二层平面图 1：250
总平面图 1：500
经济技术指标
总用地面积 8800㎡
总建筑面积 2658㎡
成人阅览室
研究中心
图书资料中心
书库
珍藏库
商店

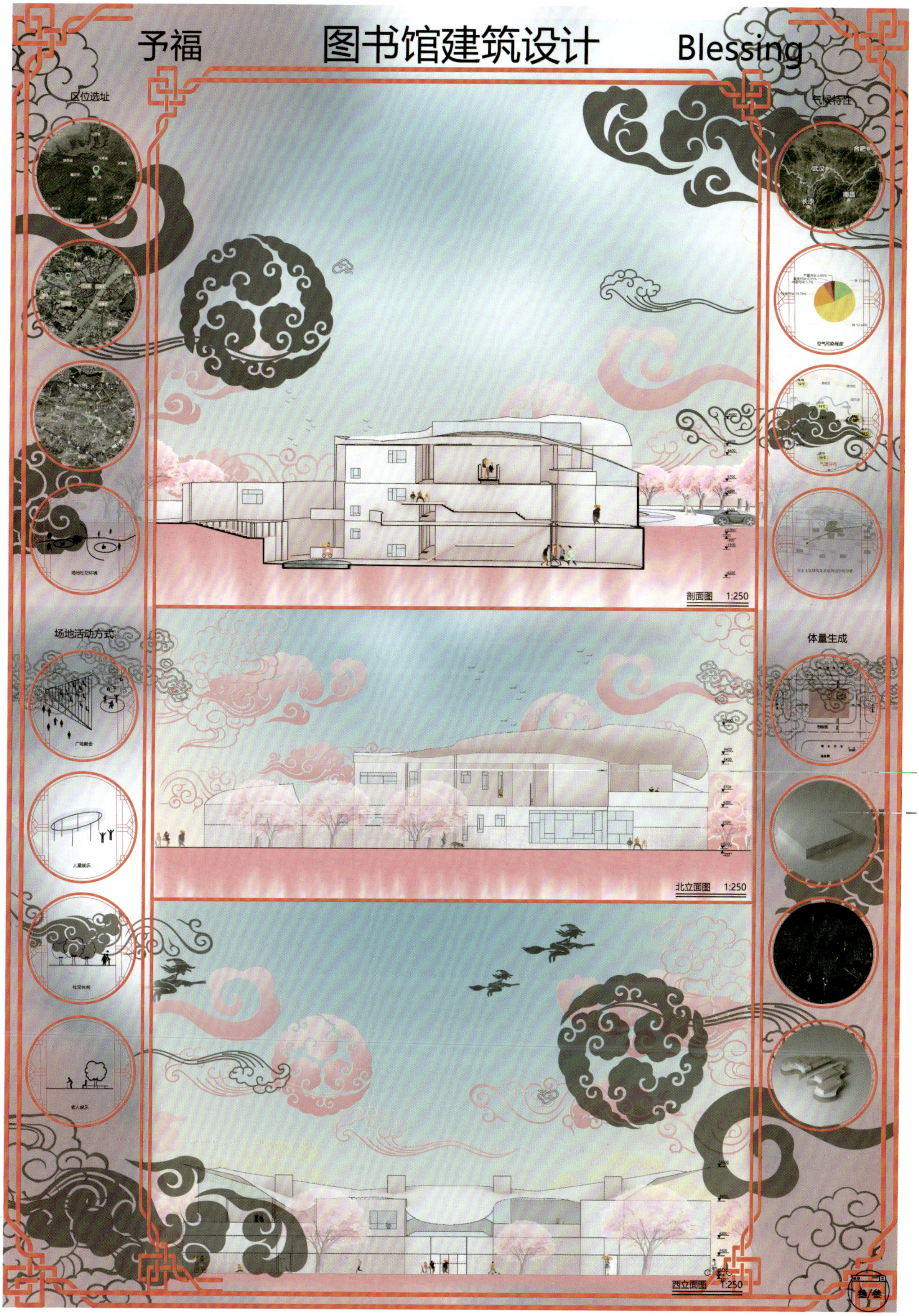
予福
图书馆建筑设计
Blessing
区位选址
气候特性
场地活动方式
体量生成
剖面图 1:250
北立面图 1:250
西立面图 1:250

■ 其他作品 Other Works

未来图书馆设计

Future Library Design

学校名称 University (College) Name
河南城建学院，中国
Henan University of Urban Construction, China

指导教师 Supervisor(s)
宁 凯 Ning Kai　　李晓东 Li Xiaodong

参赛学生 Participant(s)
杨 森 Yang Sen　　田坤鹏 Tian Kunpeng
夏楷林 Xia Kailin　　王 蕾 Wang Lei

设计说明 Design Description

基于可持续发展的城市更新：二十一世纪的城市建设正处在一个“新城代谢”的时期。传统的建筑设计忽视了城市经济、社会和生态环境的可持续发展。未来需要将可持续的理念作为未来建筑设计的指南，注重城市质量的提高，包括美化城市生态化环境，优化城市空间结构，改善城市功能结构等。

迎合“一带一路”倡议：既然是以“一带一路”倡议为主题，设计立意要与世界接轨，不仅要有本土的文化体现，还要有未来科技的体现。为了满足阅读需求，未来图书馆应当具备综合特性。一方面，应当满足当地居民对更好生活的追求；另一方面，通过图书馆来带动外围经济。

未来图书馆的理解与定位：通过划分不同使用者的年龄、心情状态、使用目的等，本设计用建筑的语言，将情感设计与空间叙事结合历史街区背景，因此不同使用者对富有戏剧性的建筑空间，对时间空间的感知含有不同的心理感受。

Urban renewal based on sustainable development: The urban construction in the 21st century is in a phase of "urban upgrading". Traditional architectural design considers little of the sustainable development of urban economy, society and ecological environment. In the future, we need to take the concept of sustainability as the guide for future architectural design and focus on improving the quality of our cities, including beautifying the urban ecological environment, improving the urban spatial structure, and improving urban functional structure.

Going with the Belt and Road Initiative: with the theme of the Belt and Road Initiative, design is meant to be in line with the world, that is, the design should reflect not only the local culture, but also the future technology. In addition to satisfying the need to read, the future library should have comprehensive functions. On the one hand, it should satisfy the local people's pursuit for a better life; on the other hand, it should drive the peripheral economy through the library.

Understanding and positioning of the future library: Users are grouped according to age, state of mind, purpose of use and other conditions, emotional design and spatial narration are combined with the historical block background through the architectural language, so that different users have different psychological feelings about the dramatic architectural space as well as a sense of time and place.

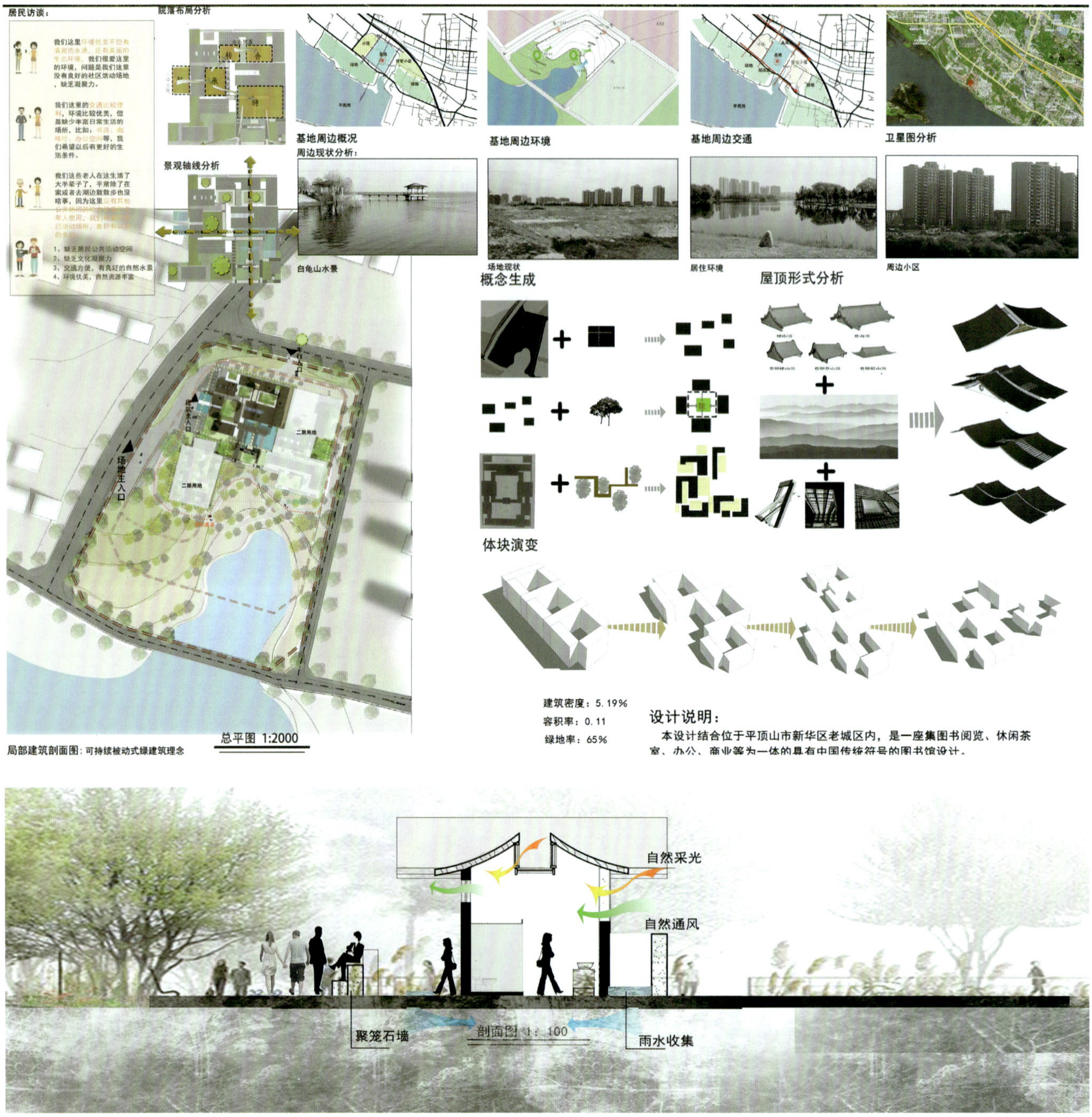
居民访谈：
我们这里环境优美不但有清澈的水源，还有美丽的生态环境，我们很爱这里的环境，问题是我们这里没有良好的社区活动场地，缺乏凝聚力。
我们这里的交通比较便利，环境比较优美，但是缺少丰富日常生活的场所，比如：书店、咖啡厅、办公空间等，我们希望以后有更好的生活条件。
我们这些老人在这生活了大半辈子了，平常除了在家或者去湖边散散步也没啥事，因为这里没有其他
1、缺乏居民公共活动空间
2、缺乏文化凝聚力
3、交通方便，有良好的自然水景
4、环境优美，自然资源丰富
院落布局分析
景观轴线分析
基地周边概况
基地周边环境
基地周边交通
卫星图分析
周边现状分析：
白龟山水景
场地现状
居住环境
周边小区
概念生成
屋顶形式分析
体块演变
场地主入口
总平图 1:2000
建筑密度：5.19%
容积率：0.11
绿地率：65%
设计说明：
本设计结合位于平顶山市新华区老城区内，是一座集图书阅览、休闲茶室、办公、商业等为一体的具有中国传统符号的图书馆设计。
局部建筑剖面图：可持续被动式绿建筑理念
自然采光
自然通风
聚笼石墙
剖面图 1：100
雨水收集

未来图书馆设计2-2
FUTURE LIBRARY DESIGN

“一带一路”建筑类大学竞赛设计

本科院校：河南城建学院　专业：建筑学
姓名：杨森、田坤鹏、夏楷林、王蕾　指导老师：宁凯、李晓东

首层平面图 1:300

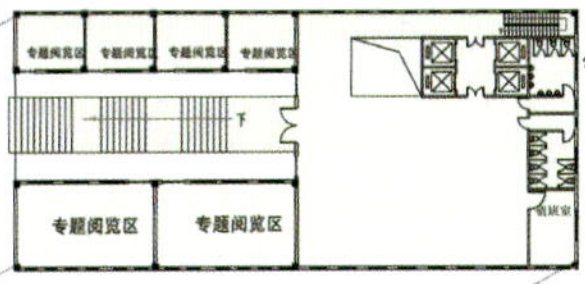

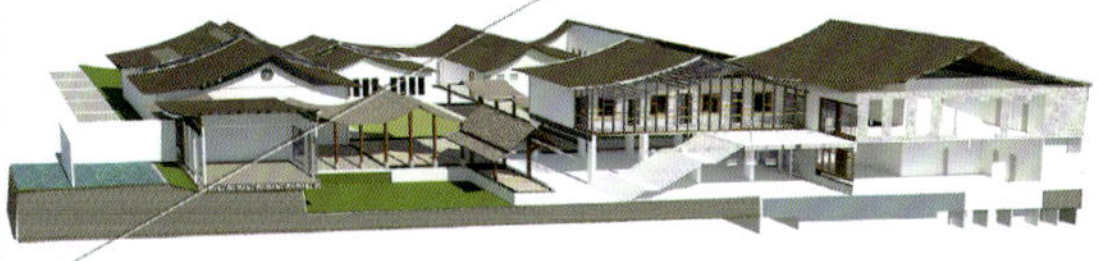

西立面图 1：300

北立面图 1：300

南立面图 1：300

■ 其他作品　Other Works

阅读 & 悦独

Enjoy Reading

学校名称 University (College) Name
吉林建筑大学，中国
Jilin Jianzhu University, China

指导教师 Supervisor(s)
张俊峰　Zhang Junfeng

参赛学生 Participant(s)
杨　璨　Yang Can
李懋凯　Li Maokai
王加琪　Wang Jiaqi
张健宁　Zhang Jianning

设计说明 Design Description

我们对于未来图书馆的设想，是在共享的立意下进行的。通过对屋顶空间的充分利用以及周围现存建筑物的功能补强，来使建筑的公共性得以加强。在使建筑空间共享的同时，与“小南湖”的环境充分融合，使居民与建筑的互动加强从阅读到“悦独”。

Our vision for the future library is based on the concept of sharing. The publicity of the building is strengthened by making full use of the various spaces of the building, including the roof, and supplementing the existing functional buildings around. As the architectural space is shared, it is fully integrated with the environment of Xiaonan Lake, so that the interaction between local residents and the building is improved.

从社会学角度看设计

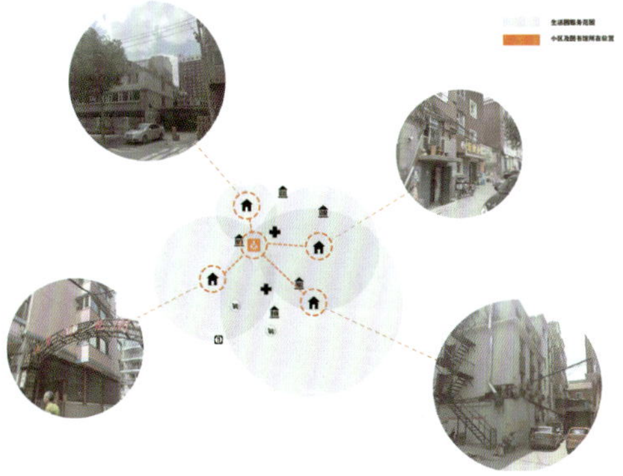

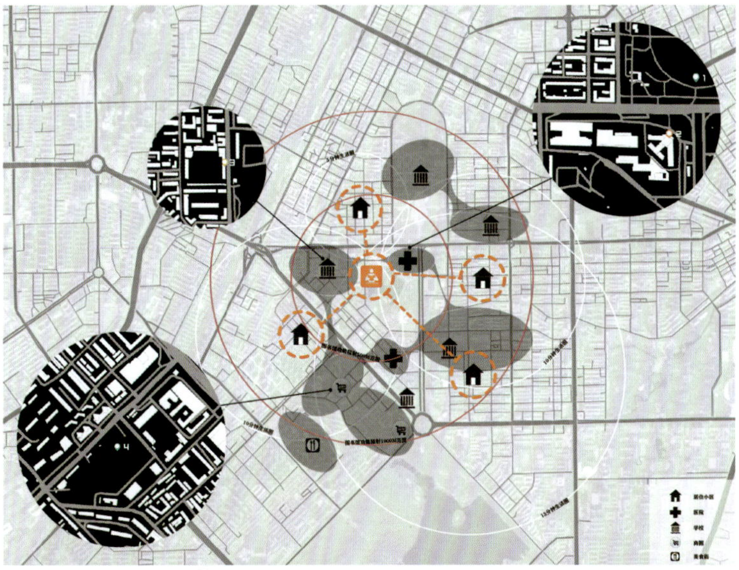

区位分析

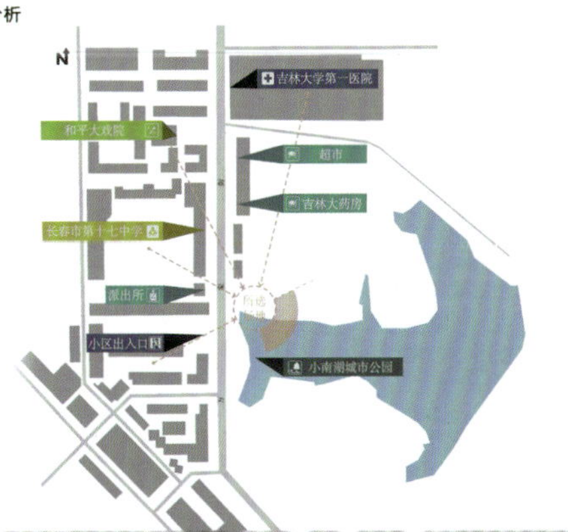

从信息数字化角度看设计（VR技术示意）

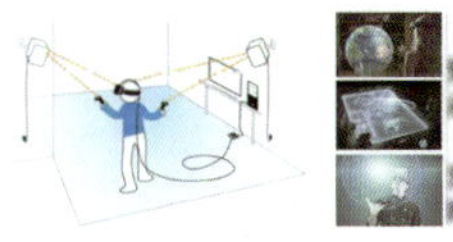

从类型学角度看设计

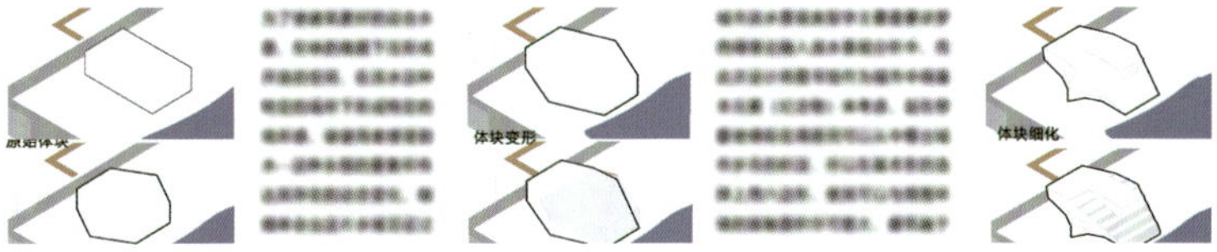

阅读&悦独

共享理念下的老街区未来图书馆设计Ⅱ

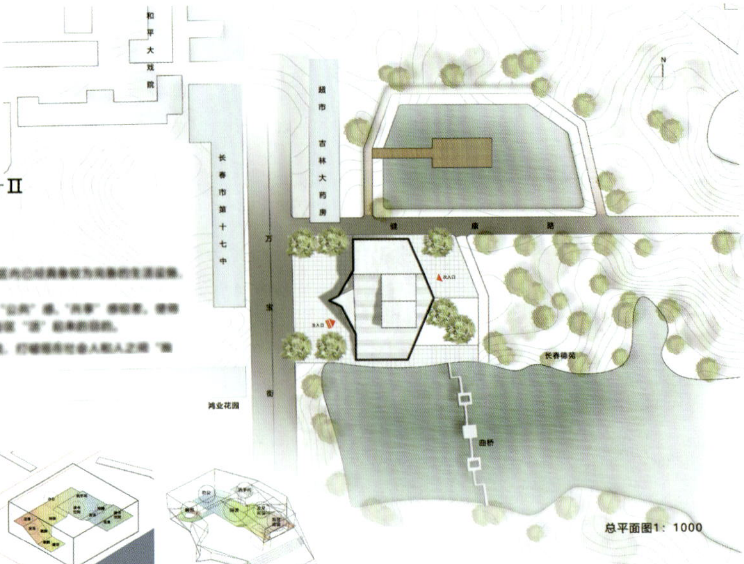

总平面图1：1000

空间设想

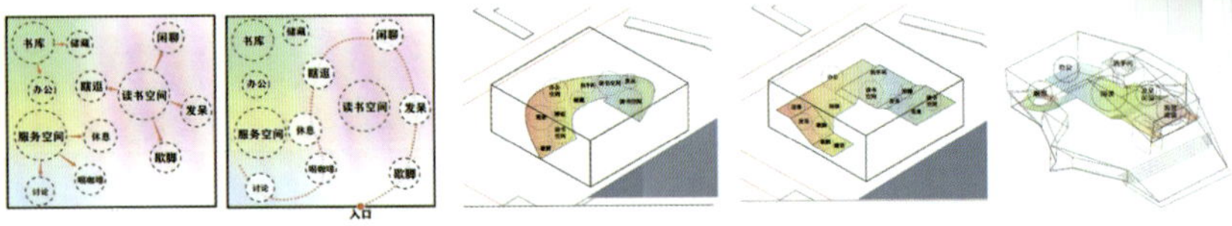

空间营造

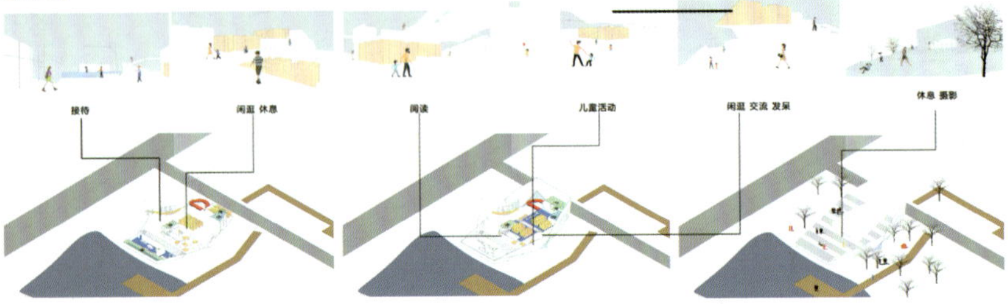

空调系统示

VR设备示意

轮椅便携装置示意

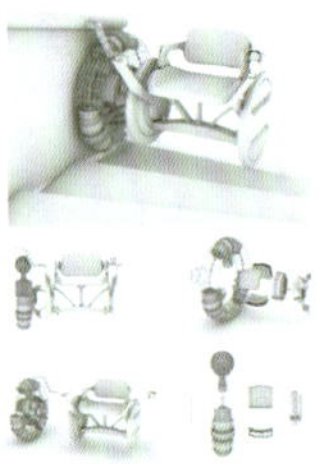

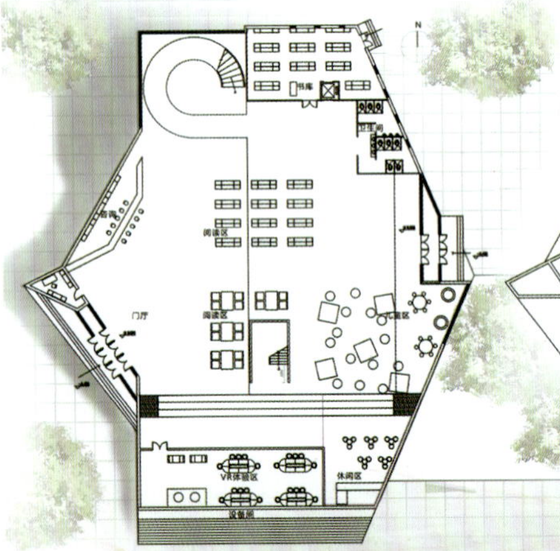

一层平面图 1：300

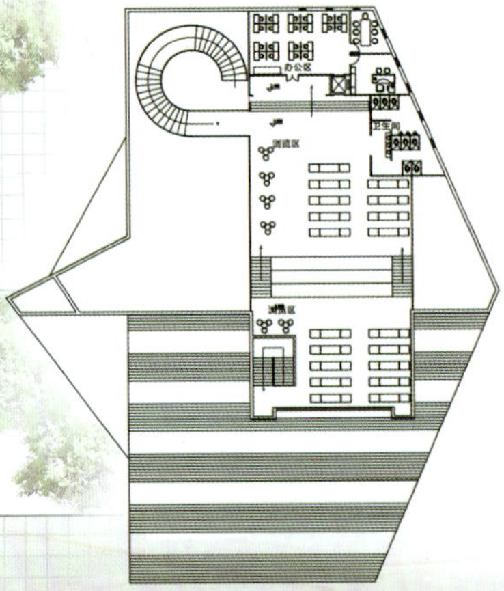

二层平面图 1：300

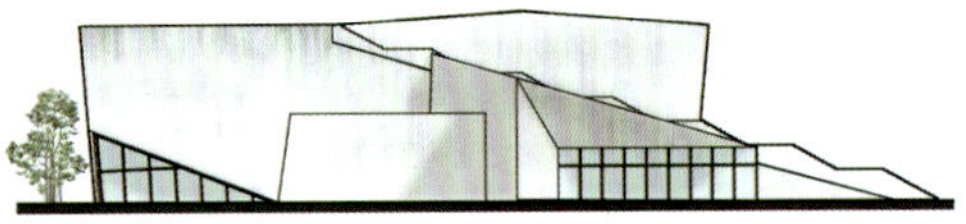
东立面图 1：300

北立面图 1：300

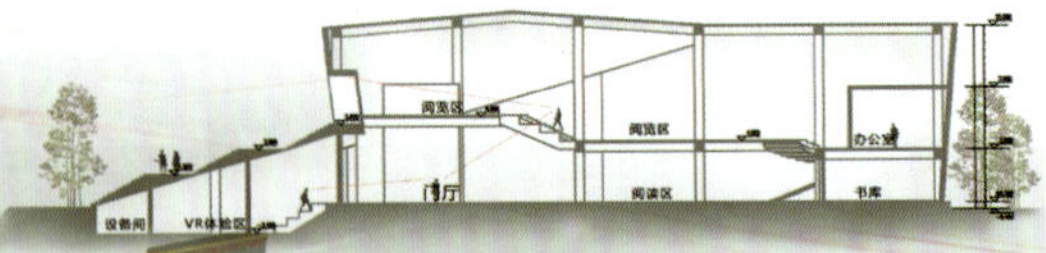

1-1剖面图 1：300

■ 其他作品 Other Works

驿·舍

Post House

学校名称 University (College) Name
吉林建筑科技学院，中国
Jilin University of Architecture and Technology, China

指导教师 Supervisor(s)
王 汉 Wang Han　　孙 佳 Sun Jia

参赛学生 Participant(s)
雷国庆 Lei Guoqing　　马博宬 Ma Bocheng
张大智 Zhang Dazhi　　张迪峰 Zhang Difeng

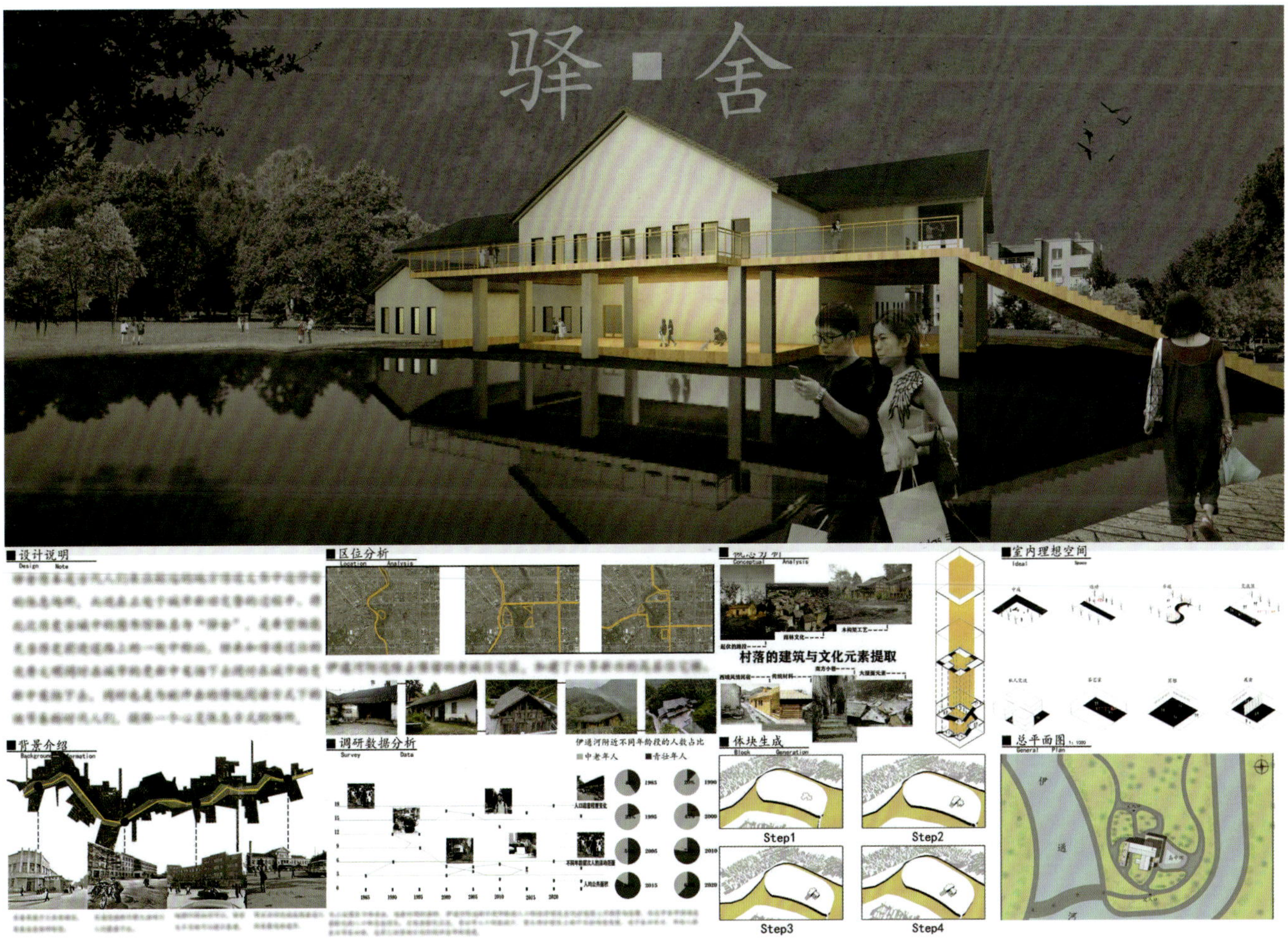

设计说明 Design Description

驿舍原本是古代人们往来较远的地方以及传送文书公文中途停留的休息场所，而现在正处于城市新旧交替的过程之中。将此次历史古城中的“未来图书馆”取名为驿舍，是希望它能充当历史前进道路上的一处中转站，继承和传递过往的优秀文明，同时在城市的更新中发扬、传承下去。同时也是为被冲击的传统阅读方式下的快节奏的时代中的人们，提供一个心灵休息放松的场所，一处生活当中的驿舍。

A post house was originally a resting place for people to stop during the delivery of letters and documents between faraway places in ancient times, now we are in the process of transition from the old to the new. The “Future Library” in this historic city is named “Post House” during the design, hoping that it may serve as a transit station on the path of history to inherit and pass on the excellent civilization of the past during the city’s renewal, at the same time, as traditional reading style is under the impact, it is also expected to provide people living in the fast-paced era a place for spiritual rest and relaxation, like a post house in life.

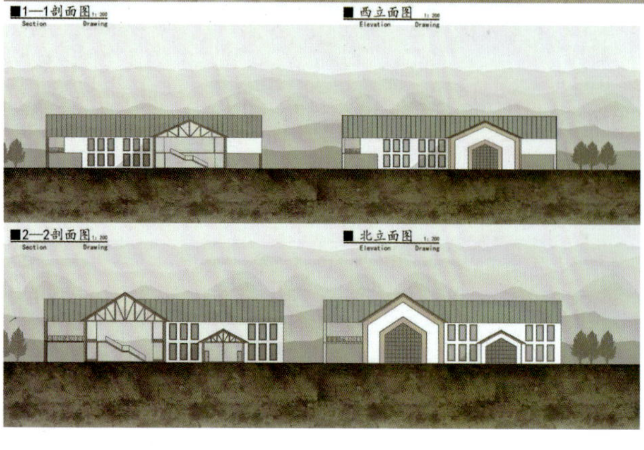

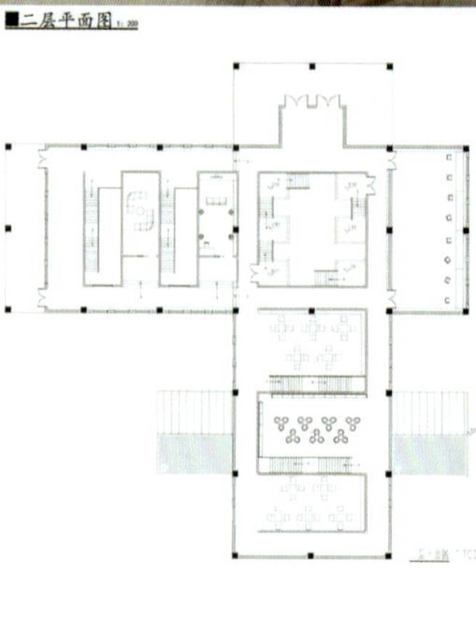

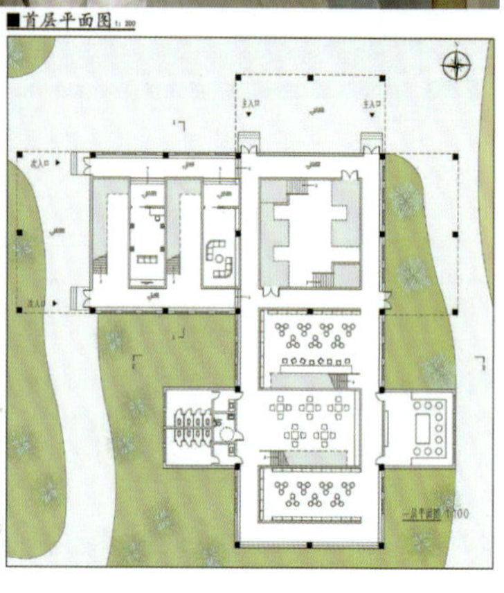

■ 其他作品 Other Works

品读 · 满铁 · 印记

Reading • Manchurian Railway • Imprint

学校名称 University (College) Name
吉林建筑科技学院，中国
Jilin University of Architecture and Technology, China

指导教师 Supervisor(s)
姜　楠　Jiang Nan　　赵　妍　Zhao Yan

参赛学生 Participant(s)
高　蕾　Gao Lei　　秦　瑛　Qin Ying
石贺文　Shi Hewen

设计说明 Design Description

长春“满铁附属地”历史街区一部分近代时期建筑保留完整，一部分已经拆除。本设计组成员对东南区块拆除部分建筑保留了大量历史照片，记录了历史印记。该区域图书馆带着它的历史印记展望未来。现代风格设计结合区域个性，把图书馆品读图书功能结合满铁历史印记，空间变化充满着矛盾与视觉冲击力。结合“一带一路”倡议，这将促进对外的互学互鉴、互利共赢的丝绸之路精神。

Some modern buildings in the “Manchurian Railway Attachment” historic block in Changchun has been preserved intact, and some have been demolished. Members of design group have kept a large collection of historical photos and historical evidence of the demolished buildings in the Southeast Block. This regional library will enter the future with its historical imprint. The design combines modern style with regional features, and integrates the library's reading function with the historical marks of the Manchurian Railway, and the space changes radiate full of contradictions and visual impact. Combining the Belt and Road Initiative, it will promote the Silk Road spirit of mutual learning, mutual benefit and win-win cooperation.

Future Library in Historic City

品读•满铁•印记：一

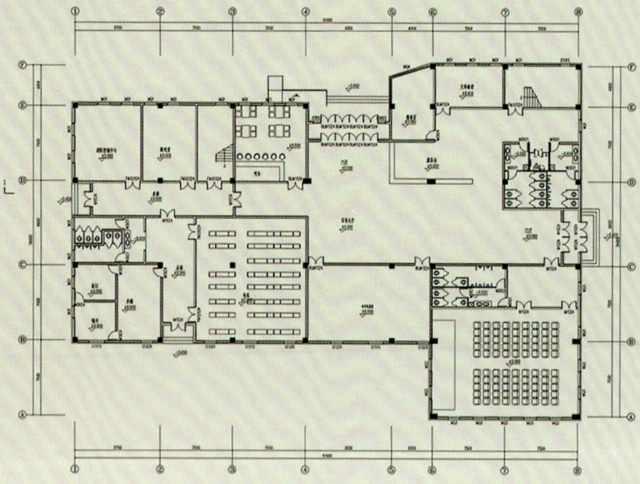

一层平面图

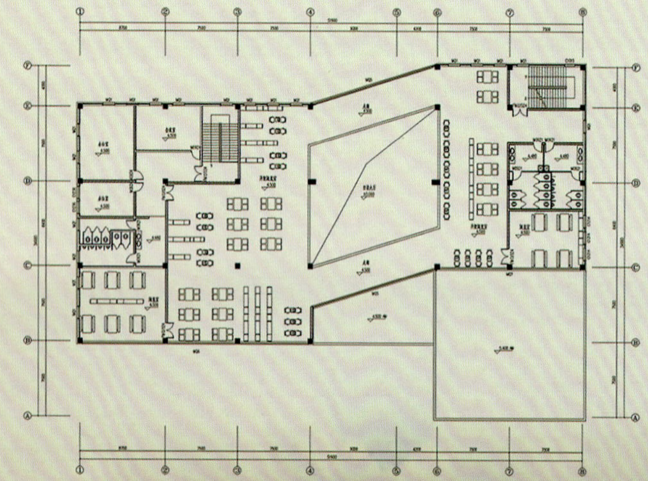

二层平面图

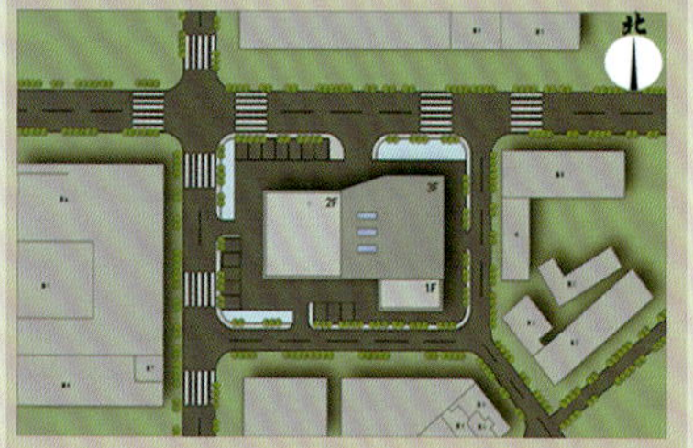

Future Library in Historic City

品读•满铁•印记：二

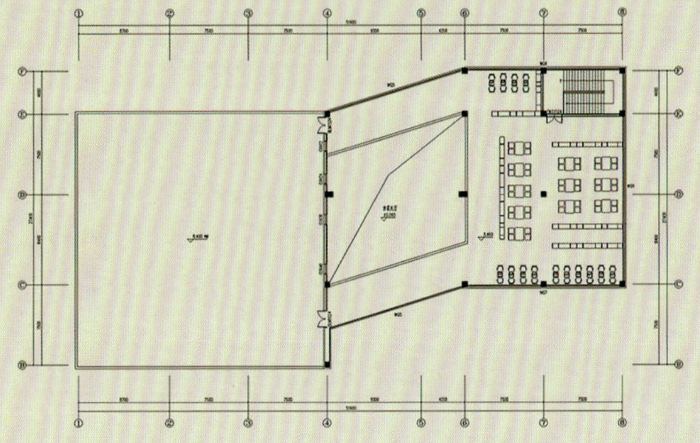

三层平面图

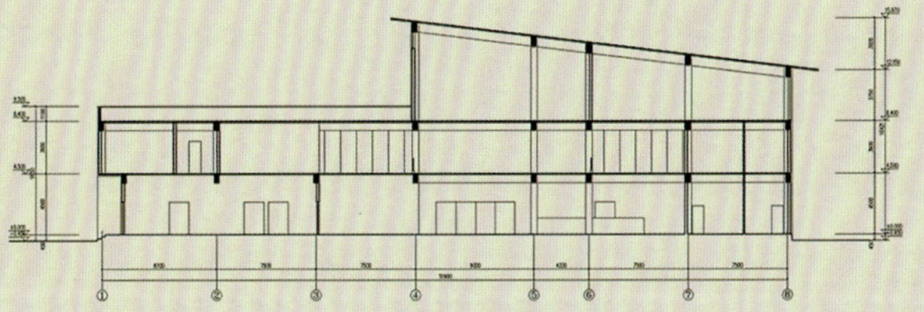

1-1剖面图

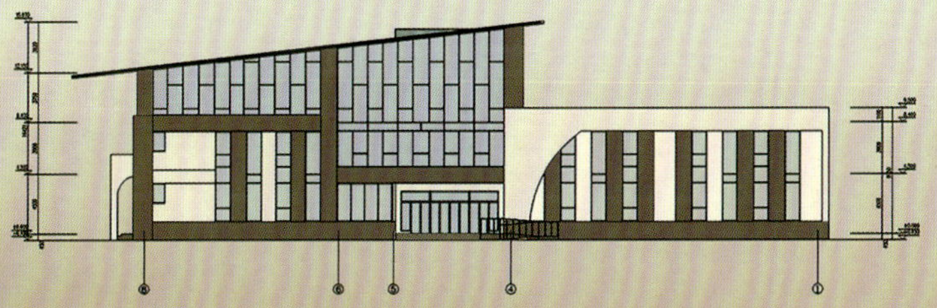

Ⓐ~①轴立面图

周边历史建筑：

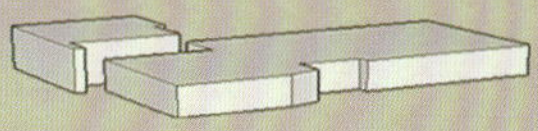

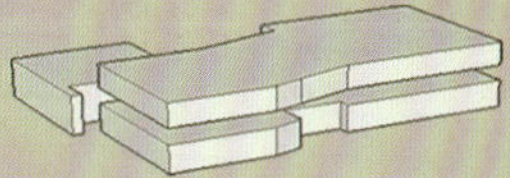

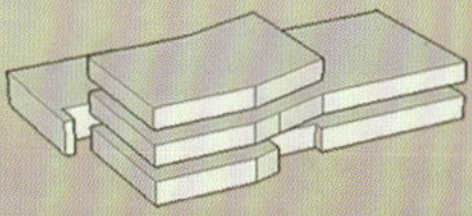

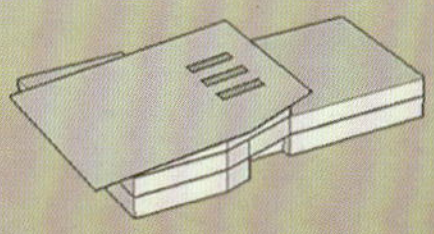

■ 其他作品 Other Works

Pustakagar

学校名称 University (College) Name
加德满都大学，尼泊尔
Kathmandu University, Nepal

指导教师 Supervisor(s)
Ar. Buddha Shrestha

参赛学生 Participant(s)
Kanchan Bastola
Kinjalk Neupane
Simran Piya
Nisha Rana Magar
Sheela Karki

设计说明 Design Description

尼泊尔博卡拉有一座200年历史的跋陀罗迦梨古庙，图书馆便位于附近，毗邻Maccha Pokhari人工湖。图书馆采用当代设计，服务于社区，欢迎各个年龄段、各种背景的人们前来一览风采。设计灵感来自“返璞归真，重返自然”的理念，形式上连接了空气、水和土地这三大自然元素，希望能调动人们的感官，号召大家去感受自然。

在科技飞速发展的时代，羁绊不再，人心远离。良好的人际关系是社会和谐的刚需，我们的设计理念便是多层面加强人际交往，鼓励交流学习。图书馆特有的互动空间和景观包含了社会和文化的本质，这样的图书馆将成为组织和构成社会的中心，不仅是信息中心，还是学习中心（艺术、音乐、职业培训等），更是娱乐和交流中心。图书馆存在的意义正在被大众遗忘，所以我们设计了这样一个图书馆，希望能重新连接人与自然、人与人。

Celebrating the historic site Bhadrakali of old Pokhara, Nepal, the library will be located near Bhadrakali old temple, which is 200 years old monument, adjacent to an artificial lake Maccha Pokhari. The library shall extend out into the community, welcoming visitors of all ages and backgrounds to experience its contemporary design. Inspired by the motto “Future follows nature”, the form connects air, water and land, the three major elements of nature evoking all the senses of visitors to experience the natural resplendence of the city.

In an age where technology is quickly advancing, human connection is being undermined. As human connection is an innate need to create a social affinity with others, our concept is enhancing its significance, encouraging the multidimensional learning process. The features of the library including interactive spaces and the landscape embraces the essence of social and cultural premises, it will be a vital center for organizing and joining society being perceived not only as an information institution but also as a learning center (arts, music, vocational training etc.), and organizer of recreation and communication. At the moment while the purpose of library is being questioned, we have envisioned a library that enriches our connection with nature and community again.

SITE BRIEF

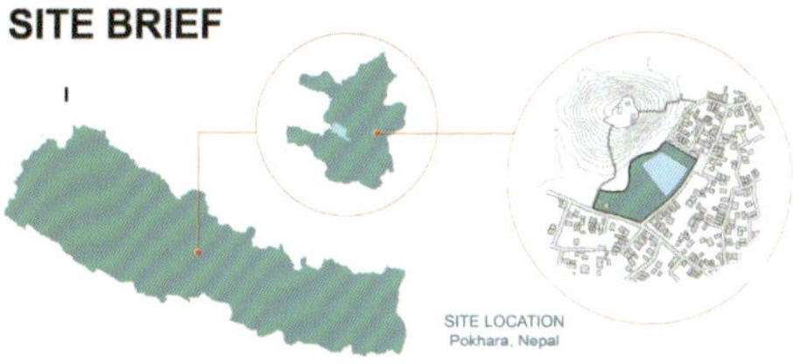

SITE ANALYSIS

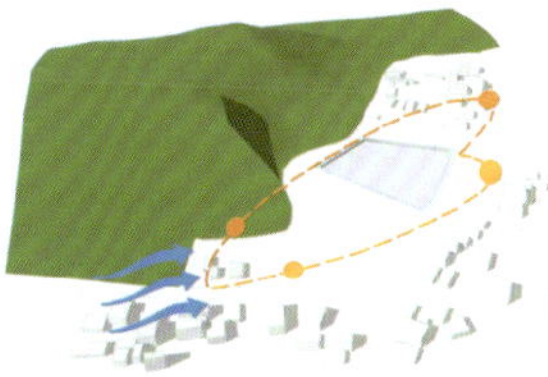

CONCEPTUAL DEVELOPMENT

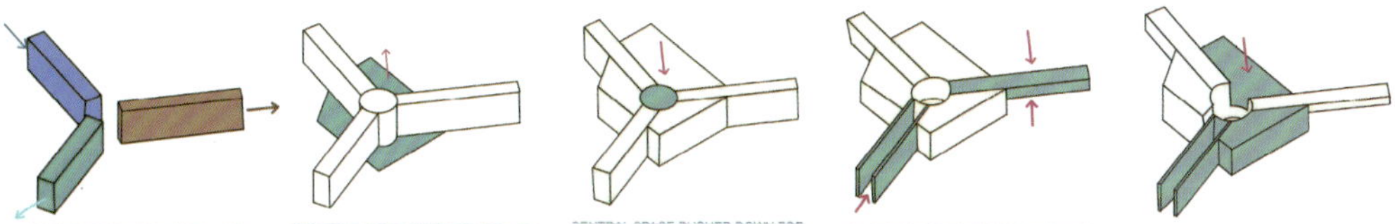

NORTHWEST VIEW OF THE BUILDING

AERIAL VIEW OF THE BUILDING

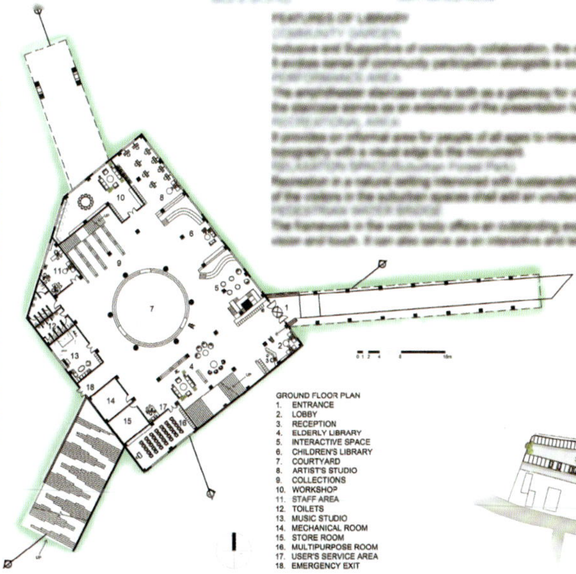

GROUND FLOOR PLAN
1. ENTRANCE
2. LOBBY
3. RECEPTION
4. ELDERLY LIBRARY
5. INTERACTIVE SPACE
6. CHILDREN'S LIBRARY
7. COURTYARD
8. ARTIST'S STUDIO
9. COLLECTIONS
10. WORKSHOP
11. STAFF AREA
12. TOILETS
13. MUSIC STUDIO
14. MECHANICAL ROOM
15. STORE ROOM
16. MULTIPURPOSE ROOM
17. USER'S SERVICE AREA
18. EMERGENCY EXIT

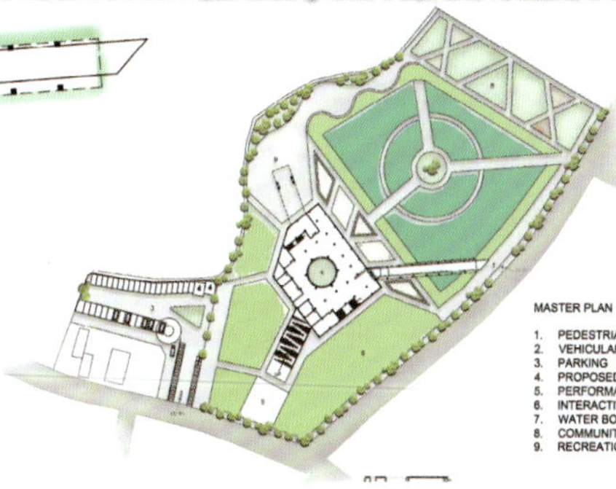

MASTER PLAN
1. PEDESTRIAN ACCESS
2. VEHICULAR ACCESS
3. PARKING
4. PROPOSED BUILDING
5. PERFORMANCE AREA
6. INTERACTIVE SPACE
7. WATER BODY
8. COMMUNITY GARDEN
9. RECREATIONAL SPACE

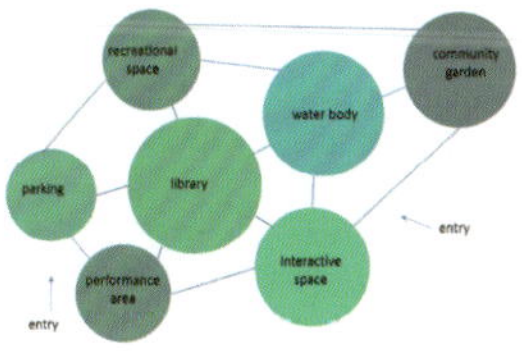

ZONING OF SPACES

EAST ELEVATION

WEST ELEVATION

PUSTAKAGAR ...the library of future

INTERIOR VIEW OF THE GREAT READING ROOM

SOUTHWEST VIEW OF BUILDING

ROOF TOP TERRACE VIEW

SOUTHEAST VIEW OF BUILDING

INTERIOR VIEW OF THE CENTRAL COURTYARD

EXTERIOR VIEW OF THE ENTRY WAY

INTERIOR VIEW OF MEDITATION SPACE

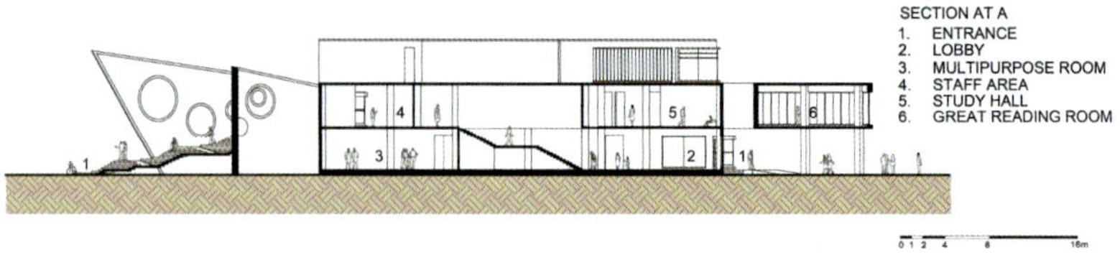

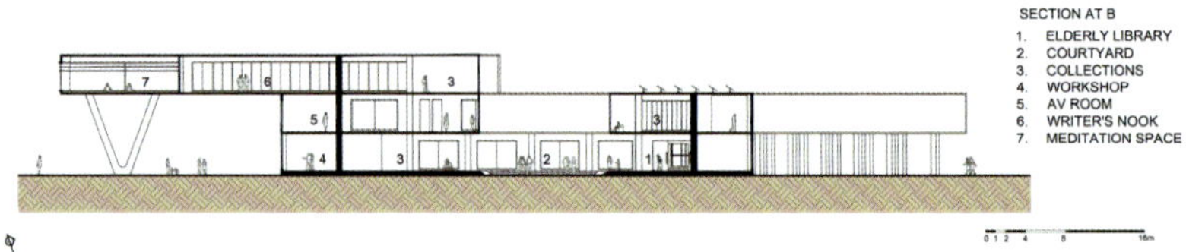

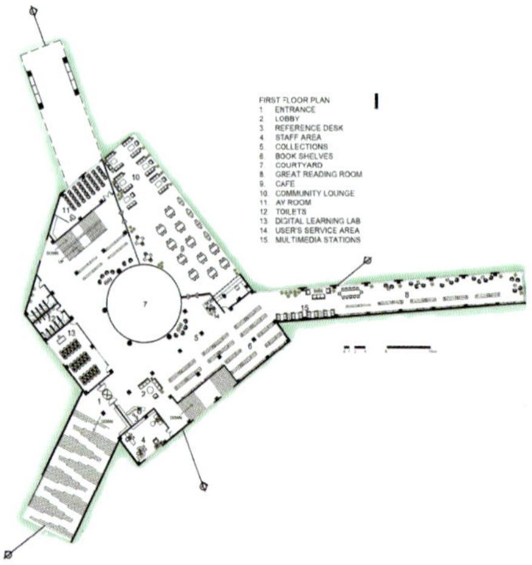

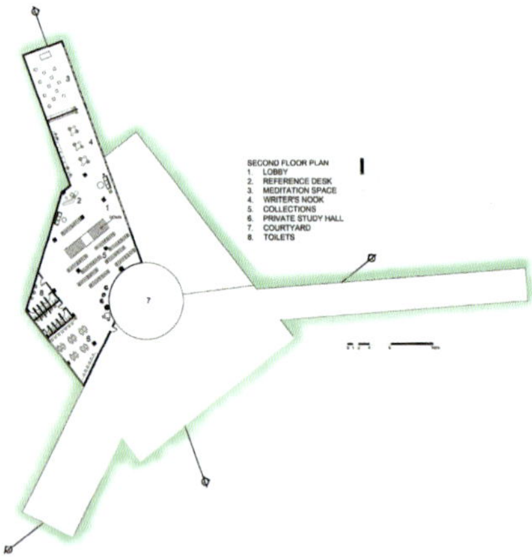

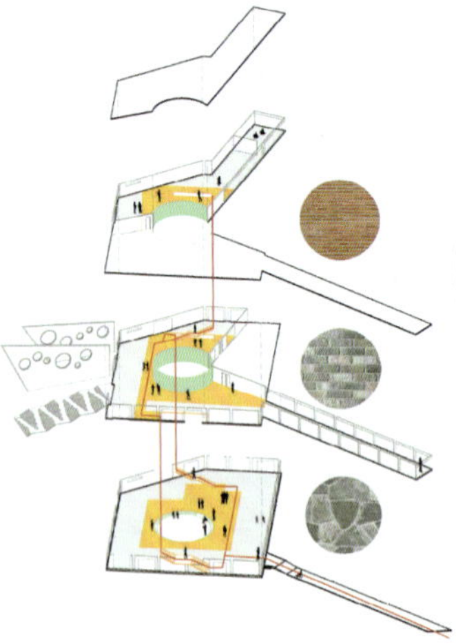
EXPLODED AXONOMETRIC VIEW

NORTH ELEVATION

SOUTH ELEVATION

PUSTAKAGAR ...the library of future

■ 其他作品 Other Works

圣彼得堡图书馆

Library in St. Petersburg

学校名称 University (College) Name
莫斯科国立建筑大学，俄罗斯
Moscow State University of Civil Engineering, Russia

指导教师 Supervisor(s)
Gazizov Timur

参赛学生 Participant(s)
Alina Krits

设计说明 Design Description

该设计方案的主题是建造一个现代化的公共图书馆建筑，但“现代化”图书馆不意味着要脱离传统。正因为这座图书馆选址在了这样一座历史底蕴丰厚的城市——圣彼得堡，它理应配得上这座城市的优雅和美丽。现代建筑与古典建筑看似冲突，但我想调和两者，设计出两者兼具的建筑。我将源自古典建筑的设计灵感与创新的建造方法相结合，最终体现于建筑的空间设计。此建筑的设计面向全年龄层，定能给所有人带去舒适、美好的体验。距圣彼得堡市中心不到 2 公里的范围内有数座岛屿，图书馆便选址于其中之一，占地约 3000m^2。它位于普里阿什卡河附近（普里阿什卡河是大涅瓦河的一条支流），附近有多所教育机构，150m 外是圣彼得堡国立交通大学的 N6 宿舍，300m 外是另外两所大学，河对岸 300m 处还有一所大学，图书馆建在这里再适合不过。此外，图书馆前是俄罗斯著名作家亚历山大・布洛克的故居博物馆这一地标，所以位置并不难找。

The idea of the design was to create a modern library building, but future can never be without past and history. Inasmuch as the building is located in an old city with a great history — St. Petersburg, it should match with the elegance and beauty of this city. Despite the difference of modern and classical architecture, I wanted to create an ensemble which would fulfil both stiles. Inspired by classical architecture, I decided to combine it with some innovative approaches to work with space. The building creates an interesting and comfortable public space for all ages and interests. The chosen location is placed on one of the islands of Saint Petersburg within 2km of the center, at the moment it is an empty land of around 3000m^2. It is located near Priashka river which flows into the Big Neva River. This location seems to be predetermined for the project, as there are multiple educational institutes nearby, which could benefit from the new Library. There is the dormitory N6 of St. Petersburg State University of Railway Transport in 150m away, there are two universities around in 300m far from the location and there is a school cross the river in 300m. Also there is the museum-flat of the famous russian writer Alexander Blok right in front, so the location will be easy to find.

F U T U R E   L I B R A R Y

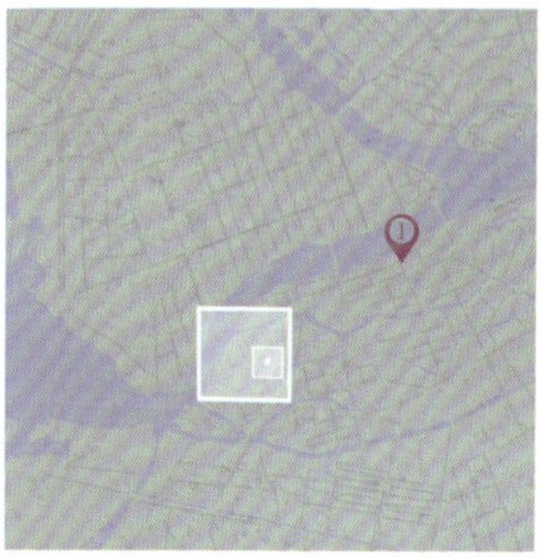

Situation plan   Scale 1:100 000

Situation plan   Scale 1:8000

Situation plan   Scale 1:3000

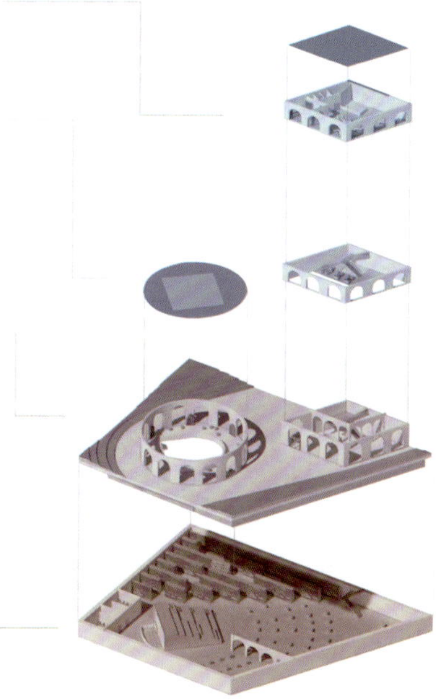

IN SAINT PETERSBURG

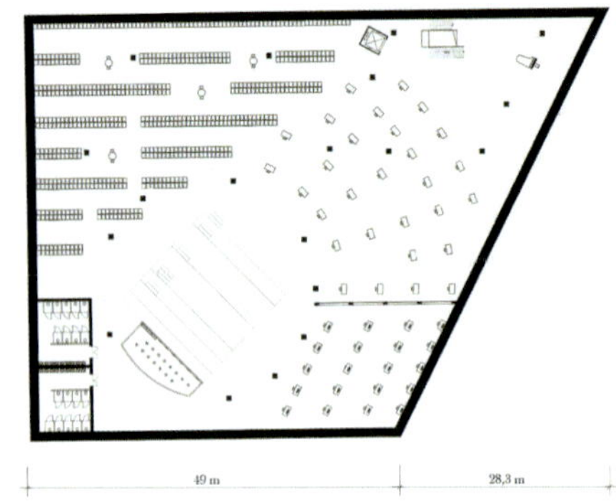

Basement floor plan Scale 1:400

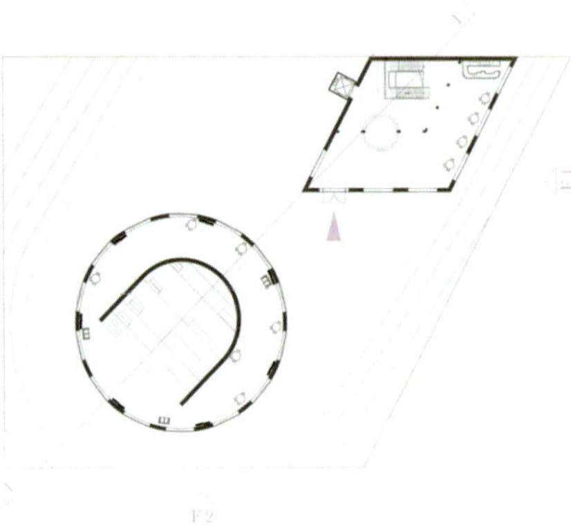

Ground floor plan Scale 1:400

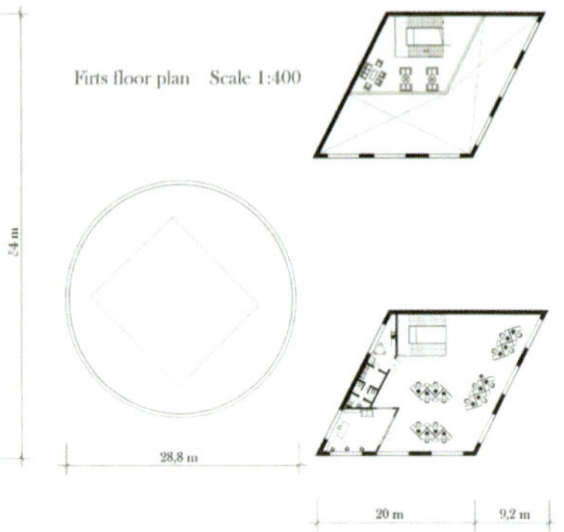

Second floor plan Scale 1:400

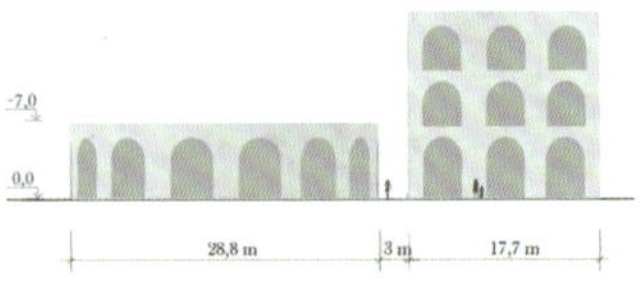

Facade 1 Scale 1:400

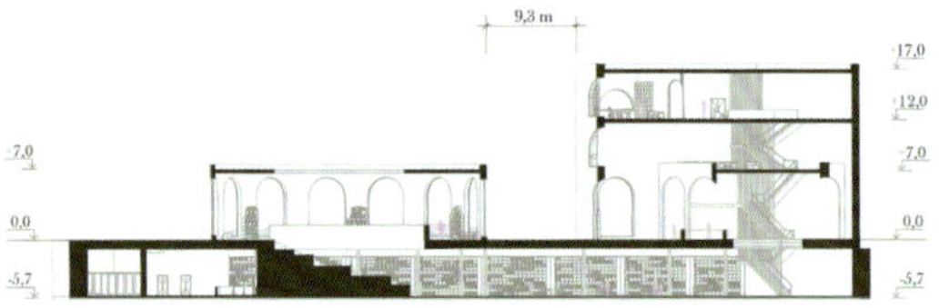

Cross section 1-1 Scale 1:400

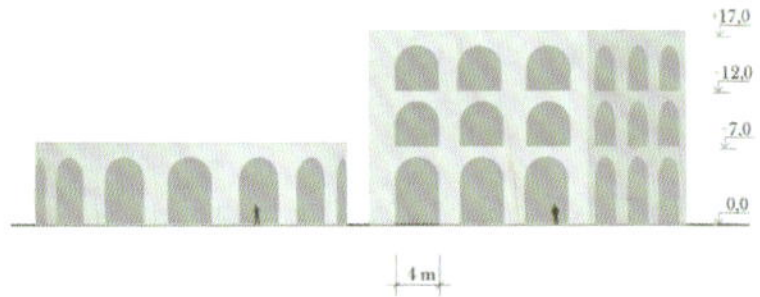

Facade 2 Scale 1:400

The Main building entrance

Relaxed reading area

Book storage and Big stairs

■ 其他作品 Other Works

“亚美尼亚圣拉扎罗岛”——图书馆改建方案 2

Saint Lazarus of the Armenians — Library Revalorisation Project 2

学校名称 University (College) Name
亚美尼亚国立建筑大学，亚美尼亚
National University of Architecture and Construction of Armenia, Armenia

指导教师 Supervisor(s)
Garegin Yeghoyan　　Raffi Bedrossian

参赛学生 Participant(s)
Lilit Soghoyan　　Ani Harutyunyan

设计说明 Design Description

首先，新建筑的外形复制了旧建筑，我们在屋顶、边角等重点位置下功夫，立面和轮廓等整体设计上尽可能模仿威尼斯的旧建筑。

根据设计，图书馆部分占地约 500m^2，属于额外新增的建筑。其中 1 号和 2 号区域各占约 300m^2。建筑一楼设咨询台和工作间，二楼设多个部分：公共档案室、历史手稿的特别储藏间、其他工作间，之所以把档案室设在二楼是出于防洪考虑，因为该地区经常发生洪灾。

关于餐厅部分，我们计划在邻近图书馆区域的上层设置玻璃餐厅，可以从楼梯直达，另外会有一个通向后厨的电梯。我们希望可以从餐厅所在的庭院看到大海，所以将餐厅设在更高、更开阔的上层，通往上层露台的通道即图书馆入口，这多少为新建筑增添了威尼斯风格。

由于目前的博物馆空间不足以容纳所有的展品，所以我们决定对其进行扩建。我们在目前内立面的基础上延伸屋角，由此获得更多空间，如您所见，新的立面大部分选用玻璃材料，可以从建筑内部看到外面的风景，增强了建筑与环境的视觉联系。同时，我们使用走廊连接新旧庭院，方便游客自由移动。

现有的观景台区域用途不大，所以我们计划将其改造并作为博物馆部分的扩展，博物馆设楼梯、庭院设坡道通向这里，我们同样采用延伸屋角和玻璃立面的方法扩大了旧观景台空间。玻璃餐厅及露台空间可以自由支配，如果天气允许，完全可以在整理后用于露天展览。

First of all, we designed a new building, the plan shape of which is drawn by continuing the already existing (old building's) lines. In the facial part, we tried to keep the existing contours to the most possible extent, by designing the roof with the similar angles as in the old building, which will be obviously seen together in the view from Venice.

The library part, which according to the task, should be 500m^2, is placed in this additional building. It is located on the 1st and 2nd, with approximately 300m^2 area each. On the first floor the consultation area and the working rooms are placed. On the second floor there are several parts — the archive consultation room at public request, special reserves for historical manuscript, and additional working rooms. We located the archive rooms on the second floor, for safety purposes in case of flood, which is not seldom in this area.

For the cafeteria, we designed a glass box in the upper level, which has access via a few stairs from the library, and there is also an elevator, which is meant to serve for the kitchen. We wanted the inner yard, which is created by adding our new building, to have more than one access to the sea, so, by taking the cafeteria box to the higher level and making openings we made it easier to get to the sea, and also used the top part the passage as a terrace with entrance from the library. This also provided some Venice style to the new constructions.

Then we decided to expand the museum part, because the area of the current space is not large enough for all the exponents. So, we kept the existing facade in the interior part and added a new one by continuing the angle of the old building's roof. As you can see, the new facade is mostly glass made, which helps to create some visual connection of the building to the environment, with beautiful views opened from the inside. Also, we thought that it would be good to connect the existing inner yard with ours, for making the movement of the visitors easier, that's why we created this corridor.

There is an existing terrace near the observatory which is not in a good condition so we decided to reconstruct it and to use it as a continuation of the museum part. You can get there not only by the indoor staircase, which is located inside of the museum, but also by using the ramp which is meant to connect it with the yard. Here we also keep the angle of the existing roof continuing it with glass panels. The space between this glass box and the open terrace is highly flexible and easy to configure and reconfigure for open air exhibition when the weather allows.

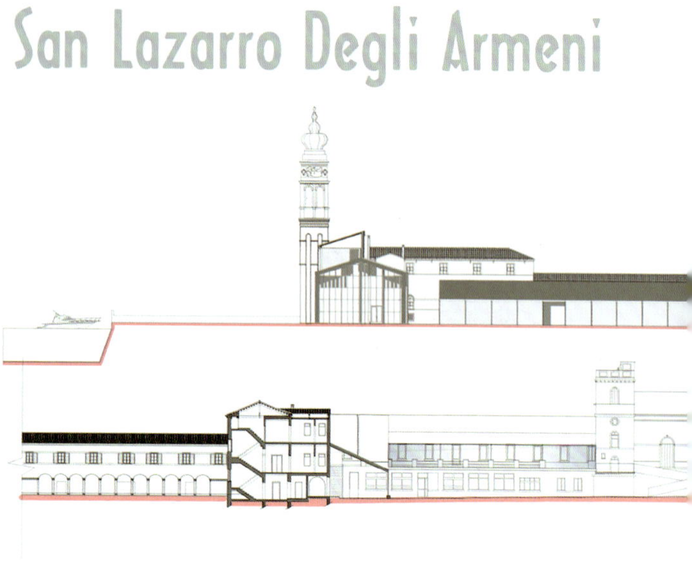

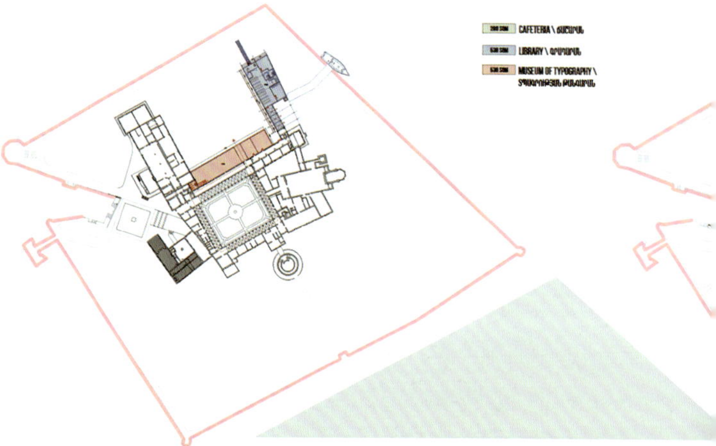

FIRST FLOOR PLAN \ ԱՌԱՋԻՆ ՀԱՐԿԻ ՀԱՏԱԿԱԳԻԾ

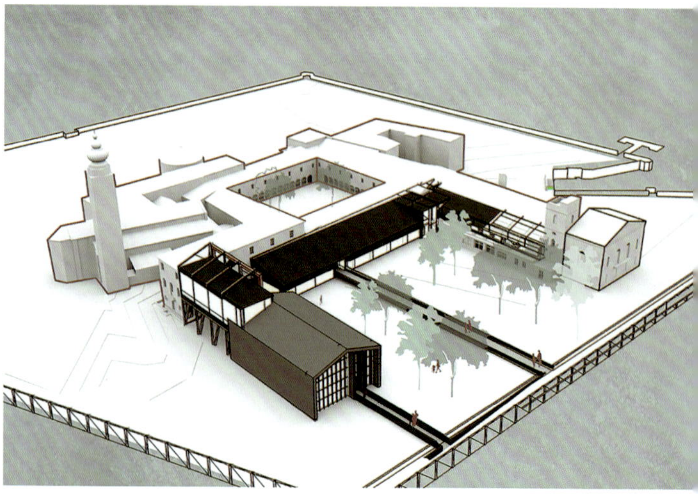

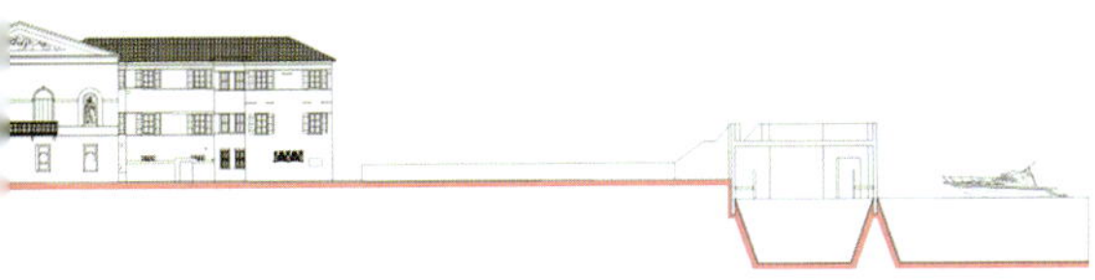

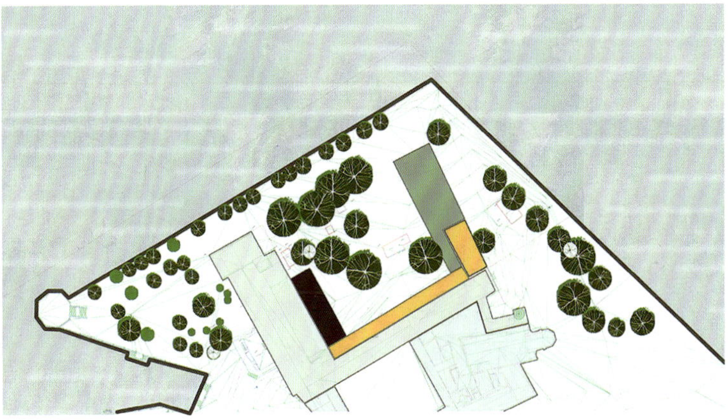

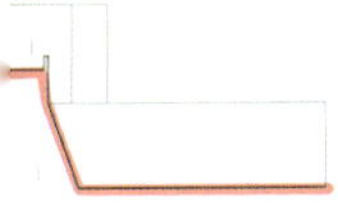

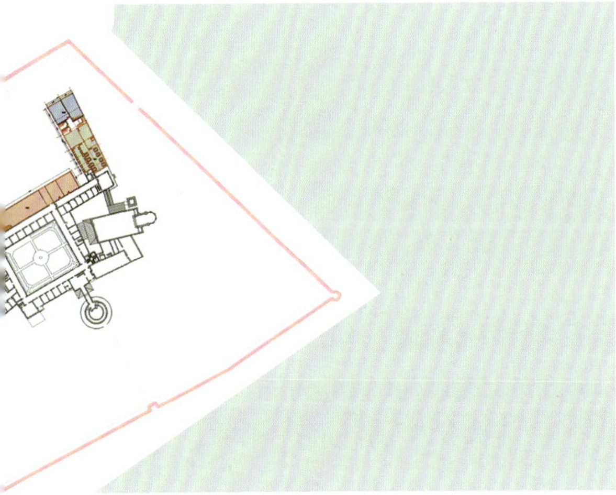

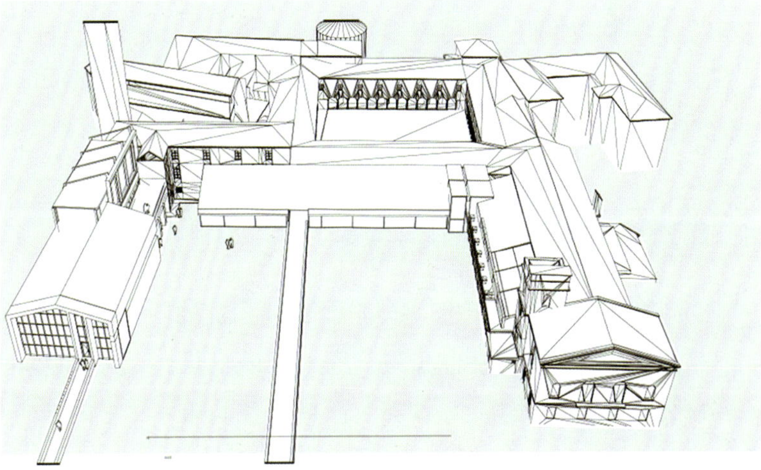

\ ԵՐԿՐՈՐԴ ՀԱՐԿԻ ՀԱՏԱԿԱԳԻԾ

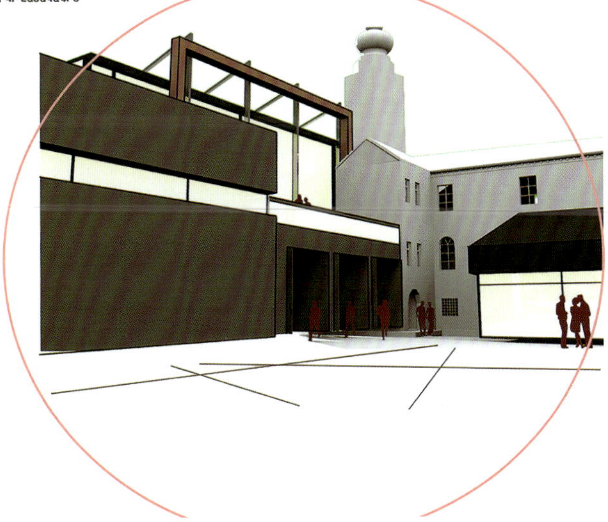

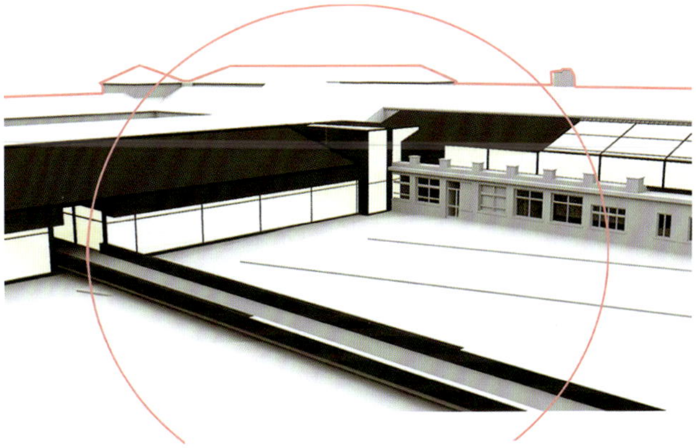

■ 其他作品 Other Works

“亚美尼亚圣拉扎罗岛”——图书馆改建方案 3

Saint Lazarus of the Armenians — Library Revalorisation Project 3

学校名称 University (College) Name
亚美尼亚国立建筑大学，亚美尼亚
National University of Architecture and Construction of Armenia, Armenia

指导教师 Supervisor(s)
Aloyan Artyom　　Sanamyan Hovhannes

参赛学生 Participant(s)
Astghik Mashinyan

设计说明 Design Description

2000 年，威尼斯和维也纳地区的 Mekhitarist 教会联合起来，统一以威尼斯 Mekhitarist 修道院为总部。它是圣拉扎罗岛的主修道院，也是维也纳的首家修道院。2002—2004 年，在威尼斯市政当局的支持下，修葺了圣拉扎罗岛的海岸、围墙、码头和广场。

我的设计也是对岛上修道院的修葺，但是更为大胆创新。在我的设计中，现存的修道院建筑会被等比例复制到近水处并调整朝向。同时，新建筑会多一个入口。至于旧建筑，我计划在已有的建筑基础上扩建其东北部。我的想法是打造新旧对比鲜明的一对建筑，在不改变岛上建筑布局前提下，为游客提供另一个入口。新建筑的屋顶形式则是参考了科学博物馆并加以创新。

由于方案包括了出版博物馆和图书馆，我计划配合亚美尼亚语字母顺序和建筑索引，依次展示 Nerses Shnorhali consisting 三十六节诗的亚美尼亚语版本和 sharakan 版本。Nerses IV the Gracious 是 1166—1173 年亚美尼亚教会的主教。他因具有博爱精神的文学创作而被称为 Shnorhali。他是神学家、诗人、作家、赞美诗作曲家，更是天主教会承认的圣人。他按亚美尼亚语字母顺序创作了三十六节诗，形式包括祷告词、礼拜歌和赞美诗。诗歌第一节以“A”开头，第二节以“B”开头，依字母顺序依此类推，并且每一节开篇和结尾的字母保持一致。原始的 sharakan 版本诗歌除了按照亚美尼亚字母排序，还有三连音押韵的效果。但英文只有 26 个字母，不同于亚美尼亚语的 36 个字母，所以翻译必须先进行重排。

旧建筑的设计也有可取之处，比如霓虹灯发光字的装饰，我准备在新建筑也加上类似的灯光装饰等，能在夜间的装饰风格与原来保持一致。

修道院的花园在游客中大受好评。某位作家于 1905 年写道："岛上……有保存完好的花圃和果园，去威尼斯旅游少不了要到圣拉扎罗岛一游，在我看来，圣拉扎罗岛的建筑与花朵缺一不可。"五月玫瑰盛放，修道士们用花瓣制作的果酱远近闻名，此外还有一种将玫瑰花瓣、白砂糖、水、柠檬汁混合调制而成叫作 Vartanush 的饮品，它取自亚美尼亚语 վարդանուշ 的发音；字面翻译是"甜蜜的玫瑰"（也是女性名字）。岛上礼品店制作和售卖的果酱销量一直很好，修道士也把它当做早餐。亚美尼亚与自然是紧密联系着的，我想在设计中尽可能加入花元素，于是我选择了"五月花（May flower）"（著名亚美尼亚画家马尔季罗斯・萨良的作品）。镶嵌画自中世纪以来就一直是威尼斯重要的艺术形式，威尼斯圣马可大教堂的整个内部处处都有精妙绝伦、金碧辉煌的镶嵌画作，托尔切洛岛上的圣母升天圣殿收藏了该地区最古老的镶嵌画，穆拉诺的 Santi Maria e Donato 教堂有 12 世纪的镶嵌画后殿和地板（1140 年）。马赛克一词来源于意大利语 mosaico，再往前是拉丁语 mosaicus，最早是希腊语 mouseios，意思是"属于缪斯女神（美神）的物品"，正好对应镶嵌画或马赛克画这一艺术形式。镶嵌画与马尔季罗斯・萨良的碰撞，想必能带来意想不到的效果。

In 2000, the Mekhitarist Congregations of Venice and Vienna were unite. Headquarters of the Mekhitarist Congregation in Venice. It is the main monastery of St. Lazar Island and the first monastery in Vienna. In 2002 — 2004, with the support of the Venice city authorities, the island's coasts, surrounding walls, docks and square were renovated.

In my proposal I repeated the cloister idea, but in the new way. I copied the same scale of the existing building and cloister, and removed it near the water and rotated it. Also I made a new entrance for visitors in the new building. I tried to keep the existing part and to create the new one in northeastern part of the site. The idea was to create a very contrast building but also something that will continue the silhouette of the island and will have another entrance for visitors. Forms of the new building repeats the scientific museum roof but in the new way.

As the new program includes Publishing museum and Library, I decided to show in each level Armenian letters and the sharakan written acrostically by Nerses Shnorhali consisting of thirty-six verses, based on the Armenian alphabet. Nerses IV the Gracious was Catholicos of Armenia from 1166 to 1173. He received the appellation Shnorhali from his contemporaries because of the very irenic quality of his writing, Nerses is remembered as a theologian, poet, writer and hymn composer. He has been recognized as a saint by the Catholic Church. He composed prayers, liturgical songs, and chants written acrostically, consisting of thirty-six verses, based on the Armenian alphabet. By this way, when the first stanza begins with "A", the next stanza begins with "B", and so on, in alphabetical order. In many cases, the letter in the beginning of each stanza is repeated at the end of the stanza. In the original sharakan, Shnorhali creates a phonic effect whereby the first letters of the lines in each triplet are identical and follow the sequence of the Armenian alphabet. The translator has used the twenty-six letters of English alphabet instead of the thirty-six letters of Armenian alphabet, and the poem has been rearranged accordingly.

I also wanted to repeat design of the old building, decorated with neon letters, so the new building to, will have these elements, and will have a similar light decoration at night.

The gardens of the monastery have been admired by

many visitors. One author wrote in 1905: "The Island [...] with its flower and fruit gardens, is so well kept that an excursion to San Lazzaro is a favorite one with all visitors to Venice. In my opinion the site is expressing many emotions not only by architecture, but also flowers." The Mechitarist monks at San Lazzaro are known for making jam from rose petal around May, when the roses are in full bloom. Besides rose petal, it contains white caster sugar, water, and lemon juice. It is called Vartanush (Western Armenian pronunciation of վարդանուշ, vardanush literally translating to "sweet rose"; also a female given name). Around five thousand jars of jam are made and sold in the gift shop in the island. Monks also eat it for breakfast. This is the deepest connection with nature, so I tried to offer to the site more "flowers" from one of the Armenian most famous painter Martiros Saryan called "May flower". Also in Venice mosaic making never went out of fashion in the Middle Ages. The whole interior of the St Mark's Basilica in Venice is clad with elaborate, golden mosaics. Other important Venetian mosaics can be found in the Cathedral of Santa Maria Assunta in Torcello from the 12th century, and in the Basilical of Santi Maria e Donato in Murano with a restored apse mosaic from the 12th century and a beautiful mosaic pavement (1140 A.D.). The word mosaic is from the Italian mosaico deriving from the Latin mosaicus and ultimately from the Greek mouseios meaning belonging to the Muses, hence artistic. I think using mosaic for creating the panting of Martiros Saryan is a way to mix 2 cultures in one and it's a way to integration.

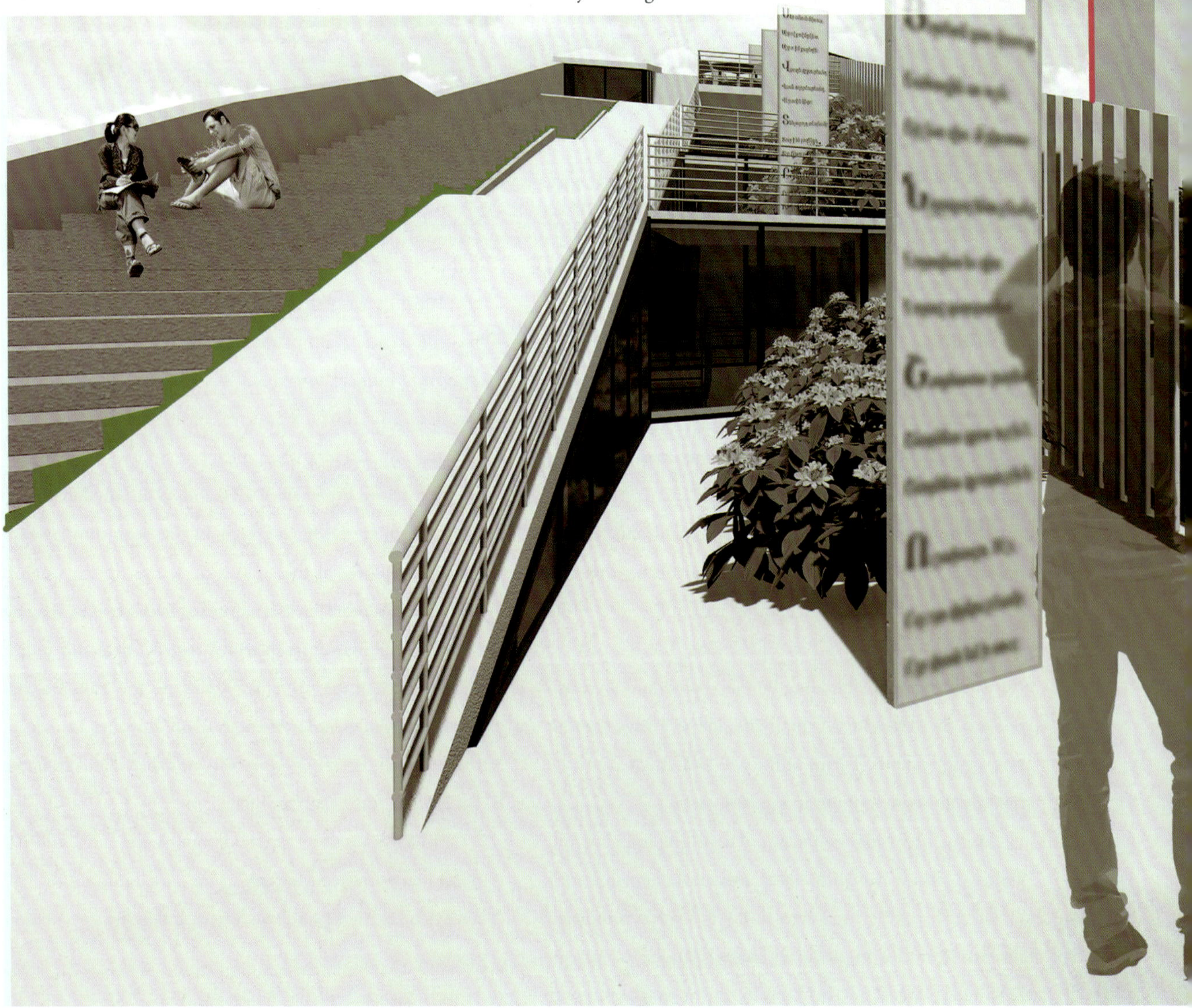

PROJECT DESCRIPTION

NEW FORM'S IDEA

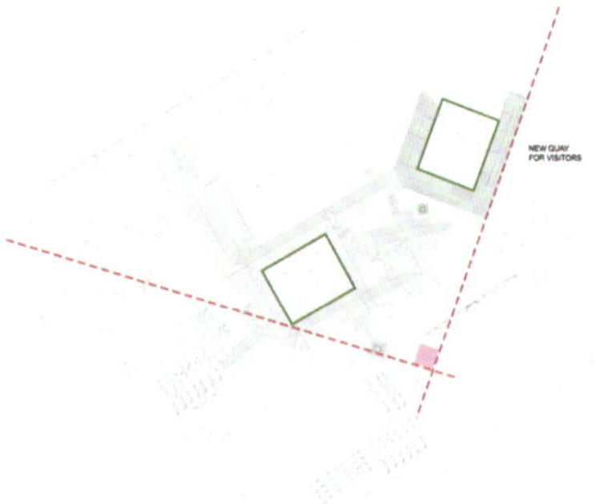

Here is one of the first sketches.

FLOOR MOSAIC

MARTIROS SARYAN
"MAY FLOWERS"

PLAQUE IDEA

MORNING OF LIGHT

■ 其他作品 Other Works

“亚美尼亚圣拉扎罗岛”——图书馆改建方案 4

Saint Lazarus of the Armenians — Library Revalorisation Project 4

学校名称 University (College) Name
亚美尼亚国立建筑大学，亚美尼亚
National University of Architecture and Construction of Armenia, Armenia

指导教师 Supervisor(s)
Emma Harutyunyan　　Garegin Yeghoyan

参赛学生 Participant(s)
Anna Karamyan

设计说明 Design Description

首先，新建筑的外形复制了旧建筑，我们在屋顶、边角等重点位置下功夫，立面和轮廓等整体设计上尽可能模仿旧建筑。

根据设计，图书馆部分占地约 500m^2，属于额外新增的建筑。其中 1 号和 2 号区域各占约 300m^2。建筑一楼设咨询台和工作间，二楼设多个部分：公共档案室、历史手稿的特别储藏间、其他工作间，之所以把档案室设在二楼是出于防洪考虑，因为该地区经常发生洪灾。

关于餐厅部分，我们计划在邻近图书馆区域的上层设置玻璃餐厅，可以从楼梯直达，另外会有一个通向后厨的电梯。我们希望可以从餐厅所在的庭院看到大海，所以将餐厅设在更高、更开阔的上层，通往上层露台的通道即图书馆入口，这多少为新建筑增添了威尼斯风格。

由于目前的博物馆空间不足以容纳所有的展品，所以我们决定对其进行扩建，我们在目前内立面的基础上延伸屋角，由此获得更多空间，如您所见，新的立面大部分选用玻璃材料，可以从建筑内部看到外面的风景，增强了建筑与环境的视觉联系。同时，我们使用走廊连接新旧庭院，方便游客自由移动。

现有的观景台区域用途不大，所以我们计划将其改造并作为博物馆部分的扩展，博物馆设楼梯、庭院设坡道通向这里，我们同样采用延伸屋角和玻璃立面的方法扩大了旧观景台空间。玻璃餐厅及露台空间可以自由支配，如果天气允许，完全可以在整理后用于露天展览。

First of all, we designed a new building, the plan shape of which is drawn by continuing the already existing (old building's) lines. In the facial part, we tried to keep the existing contours to the most possible extent, by designing the roof with the similar angles as in the old building, which will be obviously seen together in the view from Venice.

The library part, which according to the task, should be 500 sqm, is placed in this additional building. It is located on the 1st and 2nd, with approximately 300 sqm area each. On the first floor the consultation area and the working rooms are placed. On the second floor there are several parts — the archive consultation room at public request, special reserves for historical manuscript, and additional working rooms. We located the archive rooms on the second floor, for safety purposes in case of flood, which is not seldom in this area.

For the cafeteria, we designed a glass box in the upper level, which has access via a few stairs from the library, and there is also an elevator, which is meant to serve for the kitchen. We wanted the inner yard, which is created by adding our new building, to have more than one access to the sea, so, by taking the cafeteria box to the higher level and making openings we made it easier to get to the sea, and also used the top part the passage as a terrace with entrance from the library. This also provided some Venice style to the new constructions.

Then we decided to expand the museum part, because the area of the current space is not large enough for all the exponents. So, we kept the existing facade in the interior part and added a new one by continuing the angle of the old building's roof. As you can see, the new facade is mostly glass made, which helps to create some visual connection of the building to the environment, with beautiful views opened from the inside. Also, we thought that it would be good to connect the existing inner yard with ours, for making the movement of the visitors easier, that's why we created this corridor.

There is an existing terrace near the observatory which is not in a good condition so we decided to reconstruct it and to use it as a continuation of the museum part. You can get there not only by the indoor staircase, which is located inside of the museum, but also by using the ramp which is meant to connect it with the yard. Here we also keep the angle of the existing roof continuing it with glass panels. The space between this glass box and the open terrace is highly flexible and easy to configure and reconfigure for open air exhibition when the weather allows.

SAN LAZZARO DEGLI ARMENI
1717
1717-1814
1814-1890
1890-1906
1906-1949
1949-2004
EVOLUTION OF THE ISLAND BOUNDARY
PRIORITY
GENERAL PLAN
SAN LAZZARO DEGLI ARMENI
STUDENT: ANNA KARAMYAN
OTH
OSTBAYERISCHE TECHNISCHE HOCHSCHULE REGENSBURG
ensa
CLERMONT-FERRAND
IACOBUS+
01/07
2020 SAN LAZZARO DEGLI ARMENI, VENICE
CULTURE, HERITAGE AND INTEGRATION

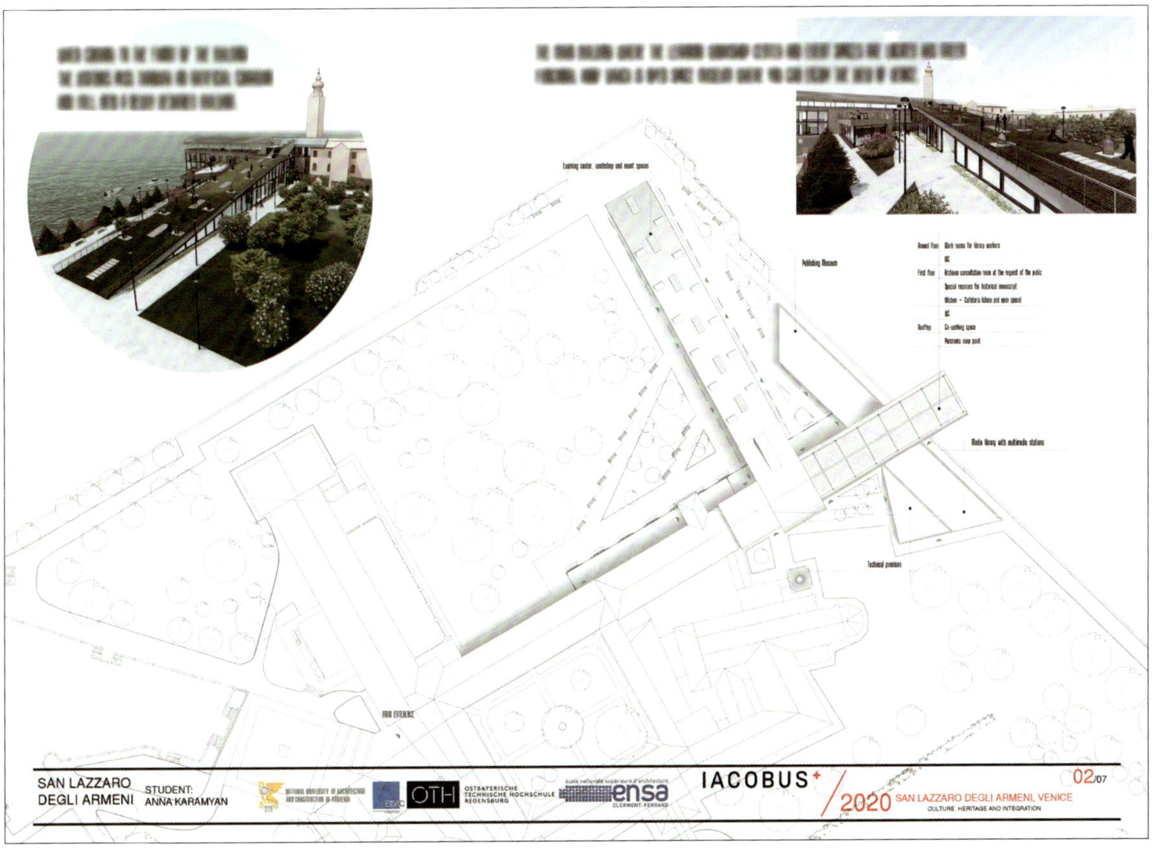
SAN LAZZARO DEGLI ARMENI
STUDENT: ANNA KARAMYAN
OTH
OSTBAYERISCHE TECHNISCHE HOCHSCHULE REGENSBURG
ensa
CLERMONT-FERRAND
IACOBUS+
02/07
2020 SAN LAZZARO DEGLI ARMENI, VENICE
CULTURE, HERITAGE AND INTEGRATION

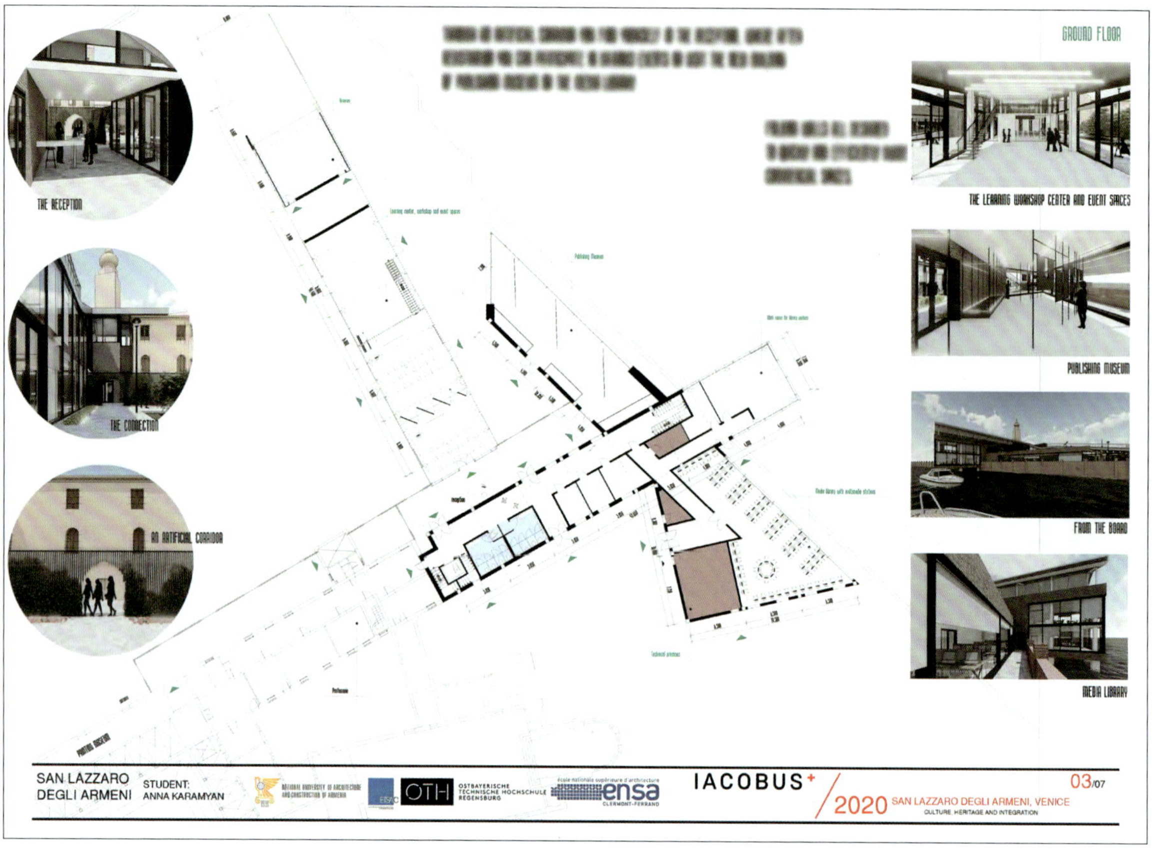
GROUND FLOOR
THE RECEPTION
THE CONNECTION
AN ARTIFICIAL CORRIDOR
THE LEARNING WORKSHOP CENTER AND EVENT SPACES
PUBLISHING MUSEUM
FROM THE BOARD
MEDIA LIBRARY
SAN LAZZARO DEGLI ARMENI
STUDENT: ANNA KARAMYAN
OTH
OSTBAYERISCHE TECHNISCHE HOCHSCHULE REGENSBURG
ensa CLERMONT-FERRAND
IACOBUS+
03/07
2020 SAN LAZZARO DEGLI ARMENI, VENICE
CULTURE, HERITAGE AND INTEGRATION

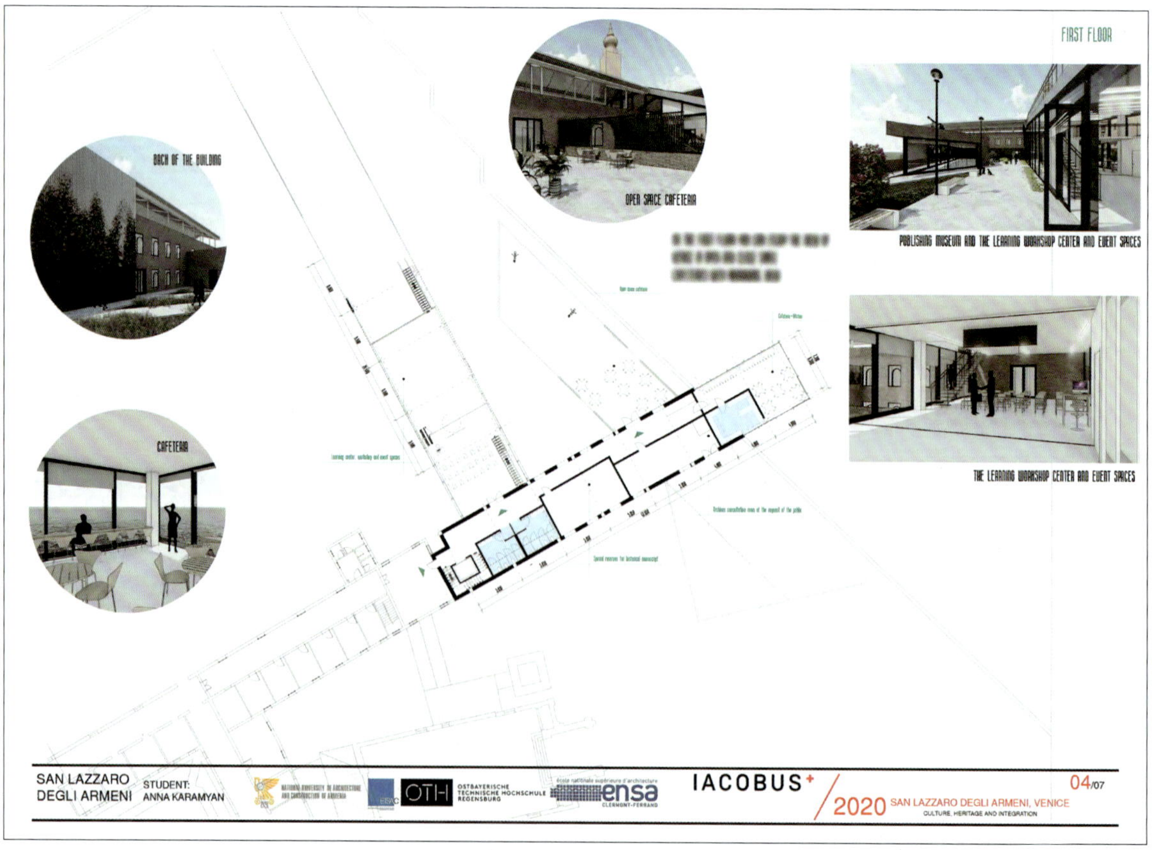
FIRST FLOOR
OPEN SPACE CAFETERIA
BACK OF THE BUILDING
CAFETERIA
PUBLISHING MUSEUM AND THE LEARNING WORKSHOP CENTER AND EVENT SPACES
THE LEARNING WORKSHOP CENTER AND EVENT SPACES
SAN LAZZARO DEGLI ARMENI
STUDENT: ANNA KARAMYAN
OTH
OSTBAYERISCHE TECHNISCHE HOCHSCHULE REGENSBURG
ensa CLERMONT-FERRAND
IACOBUS+
04/07
2020 SAN LAZZARO DEGLI ARMENI, VENICE
CULTURE, HERITAGE AND INTEGRATION

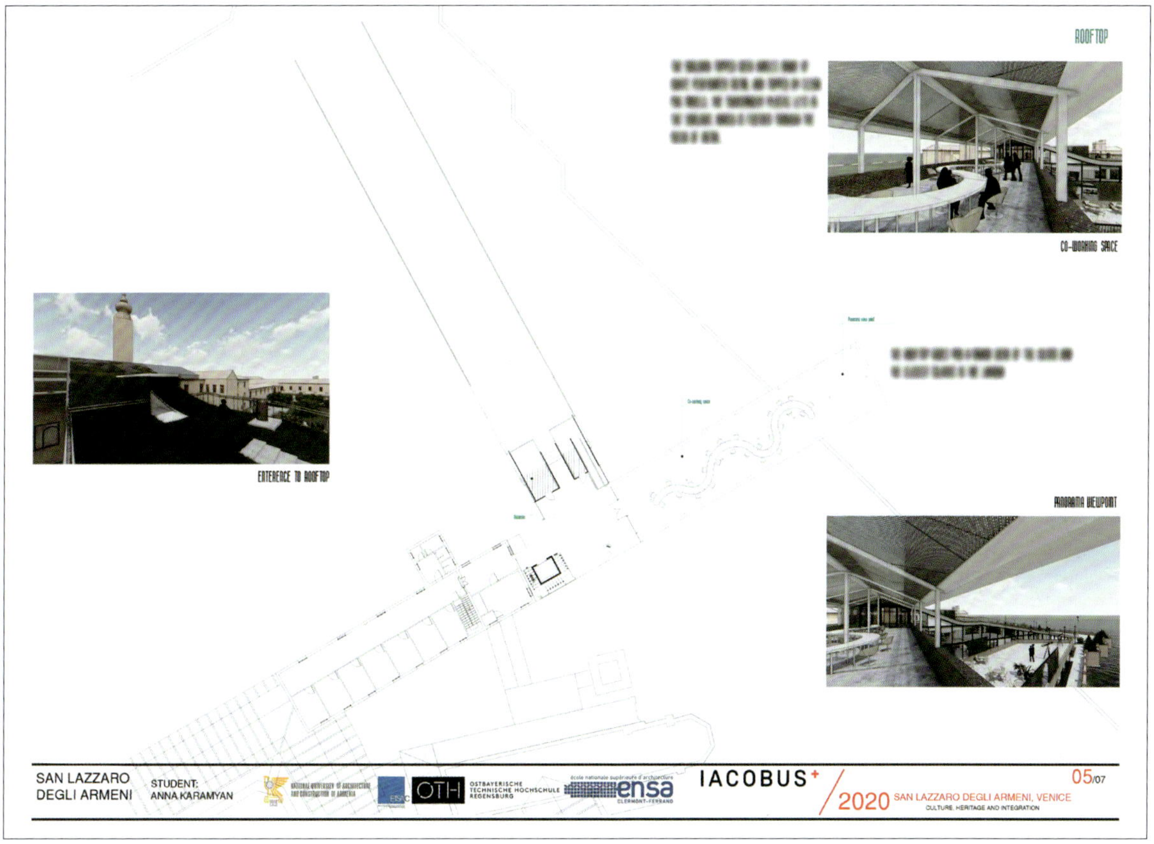
ROOFTOP
CO-WORKING SPACE
ENTERENCE TO ROOFTOP
PANORAMA VIEWPOINT
SAN LAZZARO
DEGLI ARMENI
STUDENT:
ANNA KARAMYAN
OSTBAYERISCHE TECHNISCHE HOCHSCHULE REGENSBURG
ensa
CLERMONT-FERRAND
IACOBUS+
05/07
2020 SAN LAZZARO DEGLI ARMENI, VENICE
CULTURE, HERITAGE AND INTEGRATION

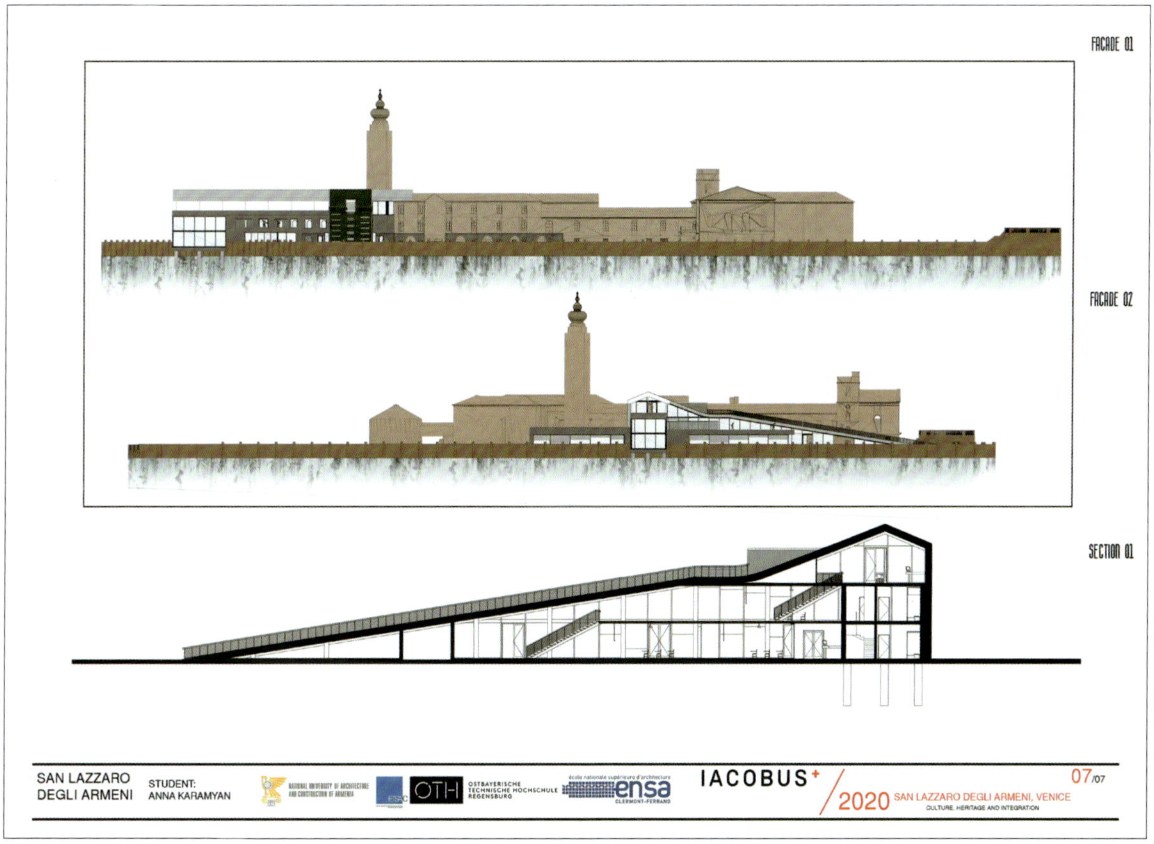
FACADE 01
FACADE 02
SECTION 01
SAN LAZZARO
DEGLI ARMENI
STUDENT:
ANNA KARAMYAN
OSTBAYERISCHE TECHNISCHE HOCHSCHULE REGENSBURG
ensa
CLERMONT-FERRAND
IACOBUS+
07/07
2020 SAN LAZZARO DEGLI ARMENI, VENICE
CULTURE, HERITAGE AND INTEGRATION

■ 其他作品　Other Works

梁赞旧城区中心图书馆

The Library Project in the Historical Center of Ryazan

学校名称　University (College) Name
莫斯科理工大学梁赞分校，俄罗斯
Ryazan Institute (branch), Moscow Polytechnic University, Russia

指导教师　Supervisor(s)
Karetnikova Svetlana Veniaminovna

参赛学生　Participant(s)
Samsonov Artem
Ershova Victoria
Chinikin Aleksey
Kartashov Andrey
Ponomarev Roman

设计说明 Design Description

该图书馆位于梁赞旧城区的特鲁别日河岸边，靠近克里姆林宫。设计的主要目标是为丰富居民生活而打造一个舒适惬意的消遣场所，该场所将全年开放。

该设计的特点体现在空间和结构的布局上。建筑围护结构尽可能使用玻璃，内部装修则采用天然环保材料。

可自由布局内部空间，图书馆分区一目了然。同时，建筑采用多层结构，上部层次分明。整个图书馆光照充足、绿意盎然，观景台还可以欣赏梁赞克里姆林宫和特鲁别日河的壮景。

人们可以在温室里读书，在被大自然的景观包围之下学习。

The library is located in the historical part of Ryazan near the Kremlin on the bank of the Trubezh River. The main idea of the design is to create a comfortable and cozy space for a pleasant pastime at any period of the year.

A feature of this design is spatial structural coverage. In building envelopes, glass is used as much as possible, finishing is carried out using natural eco-friendly materials.

The building has a free layout, with the help of transparent partitions you can divide the internal space. The architectural environment of the upper floor is multi-tiered. The whole building is filled with light and greenery. The observation deck offers a magnificent view of the Ryazan Kremlin and the Trubezh River.

Visitors will be able to read books in the conservatory, attend lectures while admiring the surrounding natural landscape.

Situational plan

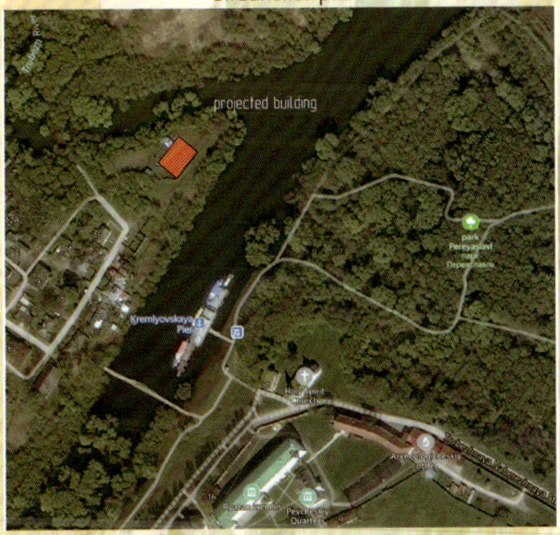

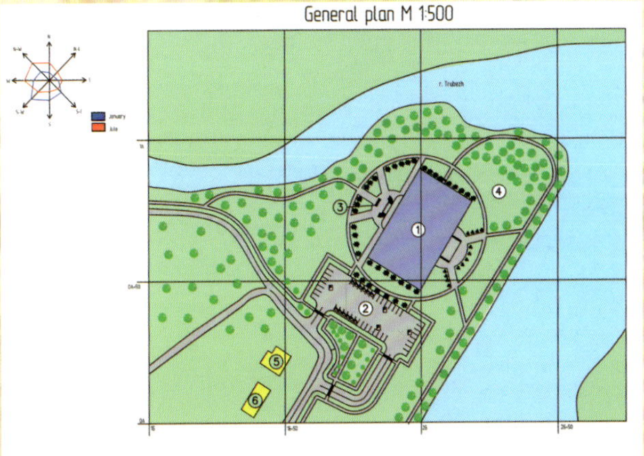

Explication of buildings and structures

№	Name	Yardage, m²	Notes
1	Projected library building	2925.00	0A+50.25
2	Parking lot for 50 cars	1875.00	0A.25
3	Fountain		0A+50.25
4	Green spaces		0A+50.25
5	Existing building	-	
6	Existing building	-	

Facade 1-14

1st floor plan

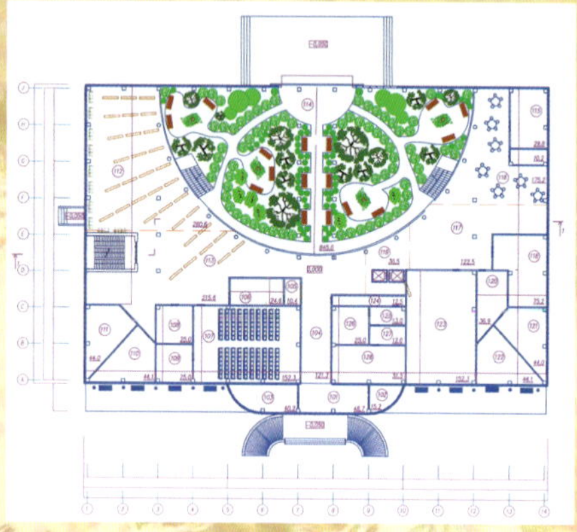

2nd floor plan

Section 1-1

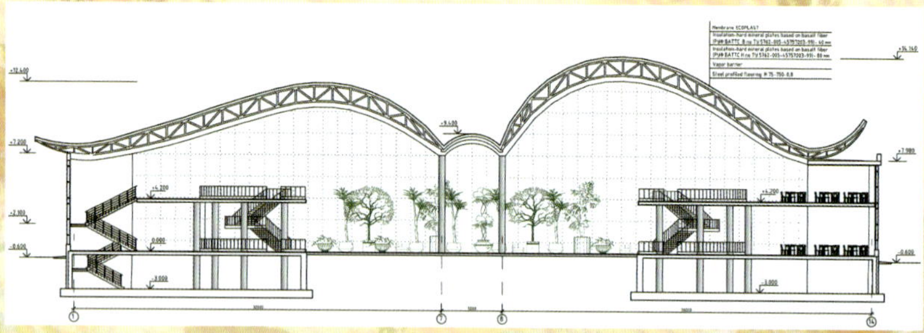

Explication of the premises of the first floor

No	Name of premises	Area, sq.m.
101	Entrance hall	49,7
102	Security room	15,2
103	Dressing room	40,2
104	Lobby	121,3
105	Information desk	18,4
106	Service room	24,6
107	Mini cinema	152,3
108	Master class zone	25,0
109	Master class zone	25,0
110	Men's room	44,1
111	Lady's room	44,0
112	Business and comfortable reading area (reading room)	280,6
113	Department for the issuance of books and subscription	215,8
114	Winter garden	845,0
115	Service premises of the coffee house	40,0
116	Coffee house	175,2
117	Mini-exhibition area	127,5
118	Children's play area	75,2
119	Elevator lobby	38,5
120	Lobby	36,9
121	Lady's room	44,0
122	Men's room	44,1
123	Internet-room	152,3
124	Lobby	12,5
125	Lobby	13,0
126	Accounts department	25,0
127	Reception director	12,0
128	Director's office	51,5

Explication of the premises of the second floor

No	Name of premises	Area, sq.m.
201	Exhibition hall	239,3
202	Conference room	190,6
203	Lobby	9,7
204	Lobby	9,7
205	Book depository	87,5
206	Men's room	44,1
207	Lady's room	44,0
208	Service room	25,1
209	Business and comfortable reading area (reading room)	373,4
210	Department for the issuance of books and subscription	227,4
211	Master class zone	53,2
212	Sky light	845,0
213	Service premises of the coffee house	40,0
214	Coffee house	175,2
215	Mini-exhibition area	238,4
216	Elevator lobby	30,5
217	Lady's room	44,0
218	Men's room	44,1
219	Musical and literary lecture hall	87,5

Basement plan

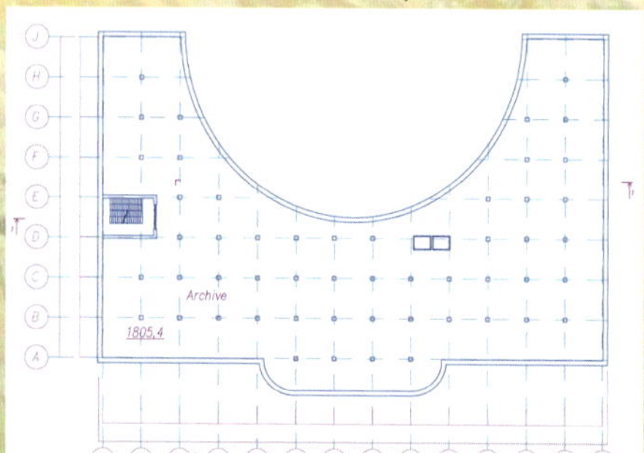

Calculation scheme in the program "SCAD 21.1"

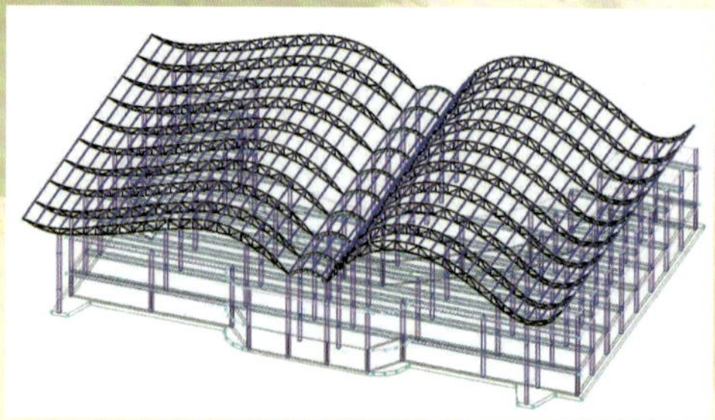

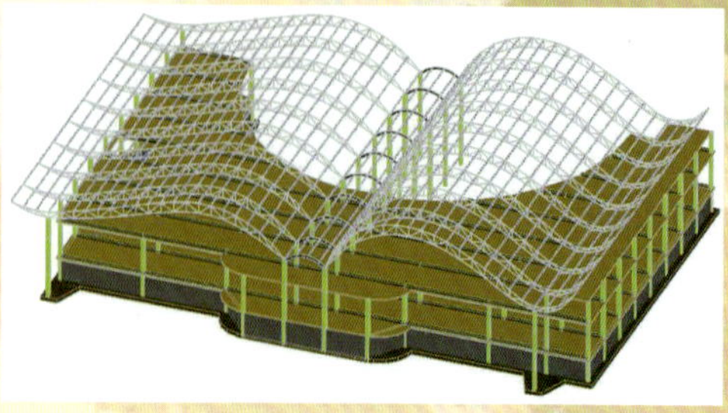

土木结构方向

Category B: Structural Design

土木结构方向——桥梁设计

我们建议学生选择合适的跨越道路路线、江河湖海、深山峡谷等地貌的桥址。可以设计公路桥、铁路桥、市政桥、景观桥、人行桥、自行车桥、综合桥等，同时需要遵循安全、适用、耐久、创新、美观、环保、节能等设计理念。

Category B (Building Structure): Bridge Design

We suggest students to select appropriate bridge sites that span road routes, rivers, lakes, seas, mountains and valleys. The bridge types include but are not limited to highway bridges, railway bridges, municipal bridges, landscape bridges, pedestrian bridges, bicycle bridges, and comprehensive bridges. In addition, each work shall follow such design concepts as safety, applicability, durability, innovation, beauty, environmental protection, energy conservation, economy, constructability, easy maintenance, etc.

■ 一等奖 Frist Prize

Salvador — Itaparica 跨海大桥

Salvador — Itaparica Bridge

学校名称 University (College) Name
长安大学，中国
Chang' an University, China

指导教师 Supervisor(s)
罗晓瑜 Luo Xiaoyu　　陈斯亮 Chen Siliang

参赛学生 Participant(s)
彭畅暄 Peng Changxuan　　Ka Thierno Seydou
高绿野 Gao Lvye　　夏心玉 Xia Xinyu
葛 恒 Ge Heng

设计说明 Design Description

Salvador — Itparica 跨海大桥计划连接巴西巴伊亚州首府 Salvador 与巴西最大的海上岛屿 Itparica，全桥总长 12.4 公里。巴西作为南美洲中我国“一带一路”倡议的重要合作国，其交通建设存在着重大的意义。我们团队由建筑与桥梁专业的学生组成。围绕巴伊亚州的招标要求，从自然环境考察、人文调研、美学设计、结构力学分析等方面完成了 Salvador — Itparica 跨海大桥的主航道双塔斜拉桥设计方案，从学生角度展现与提供一种新的设计思路，为“一带一路”倡议的建设尽一份绵薄之力。

The Salvador — Itparica Bridge is planned to connect Salvador, the capital of Brazil's Bahia, with Itparica, Brazil's largest maritime island, with a total length of 12.4 kilometers. Located in South America, Brazil is an important partner of China in promoting the Belt and Road Initiative (BRI)and its transportation construction is also of great significance. Our team is composed of students majoring in architecture and bridge. Around the bidding requirements of Bahia, the design scheme of the main channel double pylon cable-stayed bridge of Salvador — Itparica Bridge has been completed from the aspects of natural environment investigation, humanistic investigation, aesthetic design, structural mechanics analysis, etc., which shows and provides a new design idea from the perspective of students and makes a modest contribution to the construction of the BRI.

二等奖 Second Prize

链　道

Chain Road

学校名称 University (College) Name
重庆交通大学，中国
Chongqing Jiaotong University, China

指导教师 Supervisor(s)
董莉莉 Dong Lili　周水兴 Zhou Shuixing

参赛学生 Participant(s)
程　翔 Cheng Xiang　王　滔 Wang Tao
陈　力 Chen Li　常　青 Chang Qing
王　强 Wang Qiang

设计说明 Design Description

本桥取名为“链道”，寓意“一带一路”倡议的建设有如一条链带连通全球，体现相关国家之间“和平合作、开放包容、互学互鉴、互利共赢”的时代精神。

桥址选在“一带一路”倡议沿线范围内，注重将桥梁造型与景观设计及周边环境的结合，增强人在通行过程中的愉悦感，使本来枯燥交通活动变成极佳的观光体验，在无形中缩短路程，拉近彼此距离。本设计强调动态感设计和结构的动态可变设计，同时注重全时段的桥梁观感，结合灯光照明设计，兼顾夜景效果。

The bridge is named “Chain Road”, implying that the construction of the Belt and Road Initiative is like a chain connecting the world, reflecting the spirit of The Times of “peace and cooperation, openness and inclusiveness, mutual learning and mutual benefit” among relevant countries.

The site of the bridge is selected within the scope of the Belt and Road Initiative, and it pays attention to the combination of bridge modeling, landscape design and surrounding environment, so as to enhance people’s sense of pleasure in the process of passing, so as to make the originally boring traffic activities become excellent sightseeing experience, and shorten the journey virtually and narrow the distance between each other.This design emphasizes dynamic sense design and dynamic variable design of structure, at the same time pays attention to the full time bridge perception, combining with lighting design, giving consideration to night scene effect.

二等奖 Second Prize

衍 桥

Derived Bridge

学校名称 University (College) Name
沈阳建筑大学，中国
Shenyang Jianzhu University, China

指导教师 Supervisor(s)
王占飞 Wang Zhanfei

参赛学生 Participant(s)
姜乃峰 Jiang Naifeng
尚媛丽 Shang Yuanli
冯晓彤 Feng Xiaotong
杨非凡 Yang Feifan

设计说明 Design Description

本方案在规划形态上充分结合周边地理区域，紧密结合地形、场地内桥梁形态，参考周边空间机理，依托外围交通条件设置，将空间、形式、功能三者统一作为设计原则，将适用、美观、经济作为设计理念。周围中央景观区域及周围地形，展开三条人行桥流线，将空间、功能、体量有效结合。建筑形态上采用三条柔和曲线，主塔采用钢桁架结构，人行桥由可供观赏的景观和空间阶梯平台形成一个活力人行桥系统。独特的休息区，让人们在浮躁的生活中得到一丝安静。

In terms of planning form, the scheme fully combines the surrounding geographical areas, closely combines the topography and Bridge morphology in the site, refers to the surrounding space mechanism, relies on the peripheral traffic conditions, takes the unification of space, form and function as the design principle, and takes the applicability, beauty and economy as the design concept. In the surrounding central landscape area and surrounding terrain, three pedestrian bridge streamlines are developed to combine space, function and volume effectively. Three gentle curves are adopted in the architectural form, the main tower is of steel truss structure, and the pedestrian bridge has landscape and space platform for viewing, forming a dynamic pedestrian bridge system. Unique rest area, let the people living in the impetuous life get a little quiet.

■ 三等奖 Third Prize

徽路寻梦

Chasing Dreams from Huizhou

学校名称 University (College) Name
安徽建筑大学，中国
Anhui Jianzhu University, China

指导教师 Supervisor(s)
何 玮 He Wei　　石贤增 Shi Xianzeng

参赛学生 Participant(s)
常先睿 Chang Xianrui　　方远威 Fang Yuanwei
李勇健 Li Yongjian　　吕天成 Lv Tiancheng
王钱泱 Wang Qianyang

设计说明 Design Description

“徽路寻梦”是一座位于中国皖南地区齐云山国家风景区的人行景观桥。作品将马头墙、牌坊、石雕等徽派建筑特色巧妙地融入桥梁中。通过使用现代技术和材料，创新性地将传统石拱桥造型以钢筋和混凝土组合结构配合轻质 GRC 外挂装饰板的方式实现，既满足了造型要求，又满足了安全要求。采用工厂预制与现场拼装相结合的施工方法，不仅缩短了工期，也降低了对环境的影响。作品充分体现了安全、适用、耐久、创新、美观、环保、可施工等设计原则。

“Chasing Dreams from Huizhou” is a pedestrian landscape bridge which located in Qiyun Mountain National Scenic Spot in southern Anhui Province. In the work, the horse head wall, archway, stone carving and other Huizhou architectural features are skillfully integrated into the bridge. With the use of modern technology and materials, the traditional stone arch bridge is creatively rebuild in the way of steel and concrete composite structure combined with light-weight GRC hanging decorative plate, which not only meets the requirements of modeling, but also meets the safety requirements. The prefabricated construction method is adopted in the work, which not only shortens the construction period, but also reduces the impact on the environment. The work fully embodies the design principles of safety, applicability, durability, innovation, beauty, environmental protection and constructability.

徽路寻梦

■ 二等奖 Second Prize

一苇可航

One Reed can be Sailed

学校名称 University (College) Name
北京建筑大学，中国
Beijing University of Civil Engineering and Architecture, China

指导教师 Supervisor(s)
焦驰宇 Jiao Chiyu

参赛学生 Participant(s)
程冕洲 Cheng Mianzhou
赵宇昂 Zhao Yu'ang
王梦阳 Wang Mengyang
全振鹏 Quan Zhenpeng
揭　敏 Jie Min

设计说明 Design Description

雄安新区是千年大计。本桥作为雄安新区启动区核心地理位置的典型特色桥梁，为塑造启动区“中华风范、淀泊风光、创新风尚”的城市风貌，满足“中西合璧、以中为主、古今交融”的建筑特色要求。该桥位东西侧分别为城市公园绿地、广场用地和高端高新产业与科研用地。设计团队经综合考虑该桥的地理位置及周边情况，拟将该桥设计为一座“标志性”桥梁。采用无索背斜拉桥的结构形式设计，桥塔仿照具有白洋淀地域特色的植物芦苇设计，简洁大方、富有生气。桥上的“芦苇”与桥下绿洲相映成趣，共同描绘了一幅唯美画卷，彰显了雄安新区生态建设、绿色发展的决心与信心。

The Xiong'an New Area is a millennium project. As a typical characteristic bridge at the core geographic location of the Xiong'an New Area's start-up zone, this bridge is designed to shape the city's appearance combining “Chinese style, lake scenery, and innovative fashion” in the zone, and to meet the architectural feature requirements of “a combination of mainly Chinese and Western styles, ancient heritage and modern vitality”. The land lots for urban green park, square and advanced high-end industry and scientific research are distributed on the east and the west sides of the bridge. After comprehensively considering the location and surrounding conditions of the bridge, the design team plans to design this bridge as an "iconic" landmark. The structure of a cable-stayed bridge with no back-stays is adopted, and the bridge pylon is modeled after the plant reed to represent the regional characteristics of Baiyangdian, realizing simplicity, generosity and vitality at the same time. The "reeds" on the bridge create a beautiful match with the oasis under the bridge, which depicts a beautiful picture while demonstrating the determination and confidence in the ecological construction and green development of Xiong'an New Arca.

■ 三等奖 Third Prize

光明新路大桥

Guangming Xinlu Bridge

学校名称 University (College) Name
河南城建学院，中国
Henan University of Urban Construction, China

指导教师 Supervisor(s)
王　仪 Wang Yi　　屈讼昭 Qu Songzhao
董召锋 Dong Zhaofeng

参赛学生 Participant(s)
张晨宇 Zhang Chenyu　　王占岭 Wang Zhanling
郝雪东 Hao Xuedong　　丁生威 Ding Shengwei

设计说明 Design Description

本设计为预应力混凝土箱形截面连续梁桥设计，桥长为 150m，桥梁组成为 10m（桥台）+35m+60m+35m（3 跨）+10m(桥台)。桥面宽为 20m，包含 1.5m(人行道)+7m(行车道)+3m(中间带)+7m(行车道)+1.5m(人行道)。道路为城市道路，汽车荷载为 A 级。

桥梁采用预应力混凝土变高度箱形梁（A 类构件），上部主梁的施工工艺为后张法。预应力钢筋采用 1860 钢绞线，锚具采用 HVM15 型锚具，混凝土为 C50 混凝土，普通钢筋型号为 HRB400。施工方法选用悬臂施工法，上部结构的内力通过采用桥博软件进行验算。

下部结构的构件主要涵盖了盖梁、墩台、桩基、桥台等部分。其中，桥台采用埋置式 U 形桥台，桥墩采用双柱式桥墩，基础采用钻孔灌注桩基础。

The prestressed concrete box section continuous girder bridge is designed, with the bridge length of 150m. The bridge is composed of 10m (abutment)+35m+60m+35m (3-span)+10m (abutment). The bridge deck is 20m wide, including 1.5m (sidewalk)+7m (lane)+3m (intermediate belt)+7m (lane)+1.5m (sidewalk). The road is of urban road with the vehicle load of Class A.

Prestressed concrete box girder (Class A components) shall be used for bridge, and the post-tensioning method is used as the construction process of the upper girder. It is required to use 1860 steel strand as prestressed steel bars, HVM15 anchorage, C50 concrete, and HRB400 ordinary steel bars. The cantilever construction method shall be selected for construction, and the internal force of

superstructure shall be calculated by Dr. Bridge.

Substructure components mainly include bent caps, piers, pile foundation, abutments, etc. Embedded U-shaped abutment, the double-column type pier and the cast-in-place bored pile foundation are adopted.

三等奖 Third Prize

TEU 桥梁，桥下的技术元素

TEU-bridge, Technical Elements of Under-bridge

学校名称 University (College) Name
江原大学，韩国
Kangwon National University, Korea

指导教师 Supervisor(s)
Hee Chang, Eun

参赛学生 Participant(s)
Jae Hyoung An   Seung Hwan Son
Yu Sik Hong   Jun Hyuk Song

设计说明 Design Description

渔船曾是济州岛重要的海上交通工具，与岛上居民的生活息息相关，是海女们出海捕鱼和采海草的好帮手。渔船也是济州岛人民生活的产物，是海女——济州岛女潜水员的运输工具。1980—1990 年，济州岛的居民们还居住在沿海村庄的时候，家家户户都有 Teu。Teu 外形上类似大号的渔网兜，结构上有绳索连接兜环与木杆，渔民拉动木杆就能拖起渔网。Teu 的这些特点被融入桥梁设计，以展现韩国传统之美。

Jeju's traditional fishing boat was used to catch fish and to collect seaweed from the sea, it is also a product of Jeju people's lives that they used it as a means of transportation for Haenyeo, female diver in Jeju island. Until 1980 — 1990, Jeju residents still live in coastal villages with Teu prepared for each house. The structural feature of Teu is that the net is pulled up by pulling the cuddle while connecting the cuddle and the net with a cord. These Teus characteristics are applied to bridges to design bridges that express the beauty of Korean tradition.

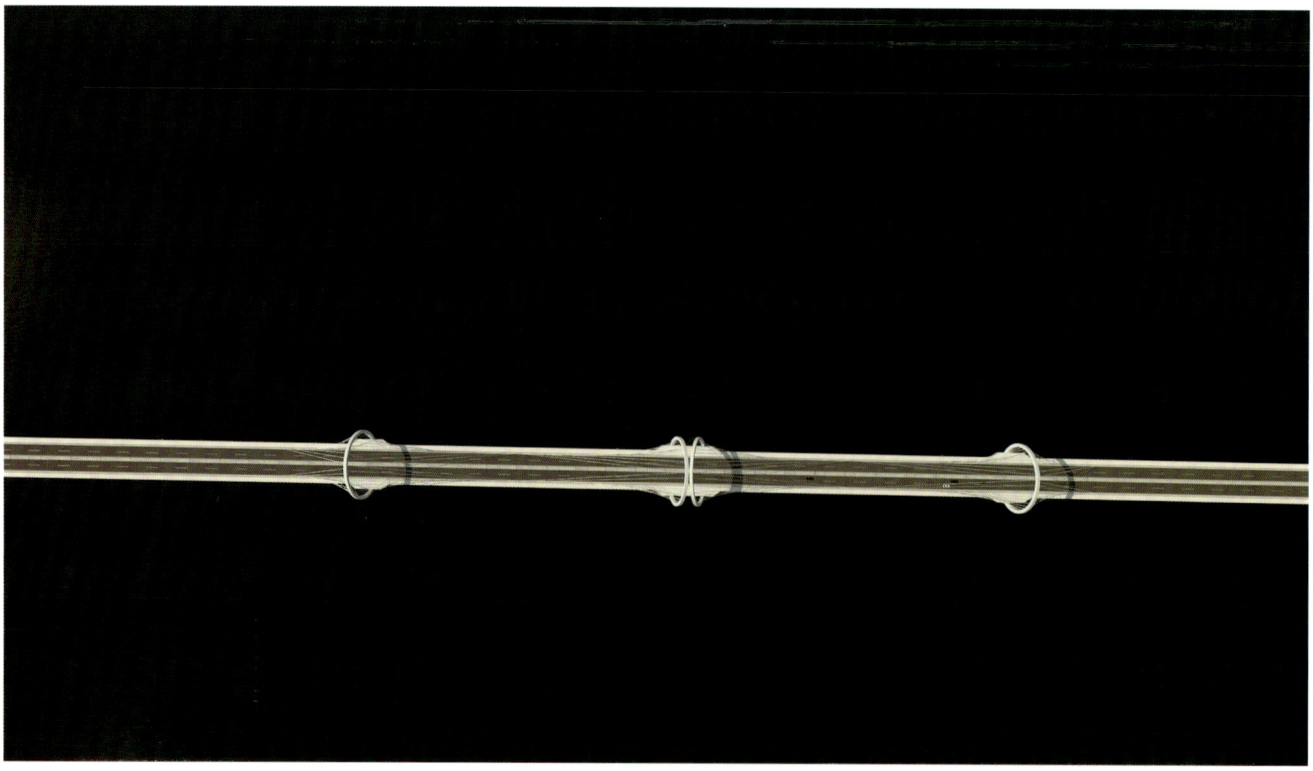

三等奖　Third Prize

横跨火星表面峡谷与火山口的桥梁
——服务于探测器及其他行星探索技术

Bridge over canyons and craters on the surface of Mars for rovers and other planetary exploration techniques

学校名称　University (College) Name
莫斯科国立建筑大学，俄罗斯
Moscow State University of Civil Engineering, Russia

指导教师　Supervisor(s)
Kolotovichev Yuri　　Smirnov Vladimir

参赛学生　Participant(s)
Svetlana Babiy　　Vitaly Zykov
Zaurbek Gadaborshev　　Valeria Maslova

设计说明 Design Description

说到土地开发，必然是离不开桥梁建设的，桥梁的功能是独一无二的。首先，人们可以通过桥梁跨越障碍，扩大交通范围。其次，两点之间线段最短，这样一条捷径可以节省能源。

在火星，桥梁建设势必会成为探索和陆地开发全过程的有效手段。

在火星建造桥梁就地取材十分重要；这样说的理由有二，一是跨星球运输混凝土和钢材的做法昂贵且低效，二是火星上的较低而又多变的温度导致常规建材性能不稳定。

在专门调查并比较了各类建筑材料后，我们得出：硫黄混凝土是理想的建材。

Bridges have always been a part of land development, the value is in its functions. Firstly, it allows people to traverse through obstacles from one place to anther. Secondly, by saving energy from crossing the obstacles, it creates the shortest seqment between two points.

On Mars, bridges construction will always be an effective solution through all stages of exploration and land development. It will allow efficient transportation of resources and vehicles. Moreover, it will open new borders and opportunities for research.

It is important to use local resources and produce construction materials on site. But there is a challenging mission: concrete and steel become excessively expensive and inefficient. More than that, the conditions on Mars make those materials vulnerable due to extremely low temperature and it's variability.

In this particular investigation the comparison of materials for construction is included. The optimal one is sulfur concrete.

■ 三等奖　Third Prize

亚美尼亚杰尔穆克地区多功能酒店桥梁

Bridge with Hotel and Multifunctional Complex in Jermuk, Armenia

学校名称　University (College) Name
亚美尼亚国立建筑大学，亚美尼亚
National University of Architecture and Construction of Armenia, Armenia

指导教师　Supervisor(s)
Hayk Asatryan

参赛学生　Participant(s)
Marieta Gevorgyan

设计说明 Design Description

本设计方案选址在亚美尼亚的杰尔穆克。这是一座热门旅游城市，所以桥梁的酒店功能可以物尽其用。

这座城市以阿帕河为界，分成两个区域。在本设计中，桥所在的路口与现代化酒店结合是一个特点，推广后能成为杰尔穆克建筑的特色。人们无须绕道到达河对岸，从路口便可直接上桥渡河，提高了出行效率。这种现代化的解决方案，不仅可以吸引游客来感受亚美尼亚宏伟的自然风光，更为当地人的生活提供了便利。

入口处看似一条带有人行道和绿化带普通的街道。令人印象深刻的是，人们驻足于桥梁之上即会成为主建造体系中最为重要的一环。桥梁采用拱门桥结构，承重部分包括 6 个坚固的拱门，和上方悬挂的其他系统。

酒店的入口旁有一个专为游客设计的大型停车场，游客可以从停车场直达酒店大厅。

酒店共8层，拥有48间客房，涵盖单人间、双人间、家庭套间、VIP 间等。

酒店第 1 层位于 0.00 位置，且与道路平齐。大厅和接待厅都在第 1 层。酒店采用垂直交通设计，游客可自由前往各楼层。

其中，SPA 中心和泳池入口都位于 –3.30 的位置。峡谷之上悬有空中泳池，并且部分使用了透明材料，游客们嬉水的同时可以游览峡谷风光。

从 –6.60 位置开始是酒店客房，其中 VIP 房间位于最后 2 层，自带悬挂于峡谷的空中阳台。

还有一个大型餐厅，入口在 +6.600 位置。这家餐厅另设有员工通道，方便运输物资，并且员工通道和游客通道互不相通。

桥的另一边是酒店的多功能区域，包括电影院、游戏区、咖啡厅、商店等入口。

为了突出建筑的游览功能，进行了专门的垂直交通设计。包括露台在内的酒店各处都设有电梯，可以直接下到峡谷，让游客亲身接触峡谷，感受美丽自然。露台上还有网球场、足球场、咖啡馆和小公园，这座桥梁将会是杰尔穆克的特色休闲场所。

The represented design is situated in Jermuk, Armenia. Considering that this resort city is very popular among tourists, a modern hotel project will be very useful.

The city is divided into two parts by the river Arpa. In this design, a new road junction is combined with a modern hotel, which can become a unique element for Jermuk's architectural composition. In fact, the junction is a new bridge for the city which gives an opportunity to reduce the time spent on going to another bank of the river. This kind of modern solutions will not only attract more tourists and let more people get acquainted with the stupendous nature of Armenia, but also make native people's life much easier.

The entrance looks like an average street with pavements, green zone etc. People feel a good impression when they step on the bridge, which also plays the biggest role in the main construction system. The bearing structure consists of 6 strong arches and the rest parts of the structural system are hung from them.

Next to the entrance of the hotel a large parking is designed for the visitors. The lobby of the hotel can be reached from the parking.

The hotel has 8 floors and 48 rooms, including single, double, family and VIP rooms.

The first floor is on the 0.00 level equal with the road. The large lobby and waiting hall are located there. Visitors can easily go to other floors due to designed vertical connections.

On the –3.30 level the SPA center is located, also, the entrance of the pool is on the same level. The pool is partly glazed and hung over the canyon, which gives a chance to the visitors to enjoy the beauty of the canyon while swimming and having fun.

The hotel rooms start from the –6.60 level. The VIP rooms are situated on the last 2 floors, have large balconies that are hung over the canyon as well.

A large restaurant is included, the entrance of which is on the mark +6.600 level. The restaurant also has different approaches from the main road for getting food, for the workers. The kitchen block and the block for workers have their own approaches from the road as well, so the visitors and the workers will not encounter.

On the other side of the bridge is the entrance of the multifunctional area of the hotel. The multifunctional center includes cinema halls, game zones, cafes, stores etc.

To become the part of the amazing nature of the city, vertical bonds are designed. There are several elevators which give an opportunity to go down to the canyon under the bridge directly from the hotel. They give a chance to visitors to get closer to the nature and to the canyon. Tennis courts, football fields, cafes and little parks are located on that terraces making them unique recreation parts for the city.

BRIDGE WITH HOTEL AND MULTIFUNCTIONAL COMPLEX IN JERMUK, ARMENIA

The represented project is situated in Jermuk, Armenia. Considering that this resort city is very popular among tourists, a modern hotel project will be very useful.

The city is divided into two parts by the river Arpa. In this project a new road junction is combined with a modern hotel, which can become a unique element for Jermuk's architectural composition. In fact, the junction is a new bridge for the city which gives an opportunity to reduce the time spent on going to another bank of the river. This kind of modern solutions will not only attract more tourists and let more people get acquainted with the stupendous nature of Armenia, but also make native people's life much easier.

The entrance looks like an average street with pavements, green zone etc. People get impression that they enter the bridge, which also plays the biggest role in the main construction system. The bearing structure consists of 6 strong arches and the rest parts of the structural system are hung from them.

Next to the entrance of the hotel a large parking is designed for the visitors. The lobby of the hotel can be reached from the parking.

The hotel has 8 floors and 48 rooms, including single, double, family and VIP rooms.

The first floor is on the 0.00 level sequal with the road. The large lobby and waiting hall are located there. Visitors can easily go to other floors due to designed vertical connections.

On the -3.30 level the SPA center is located. Also, the entrance of the pool is on the same level. The pool is partly glazed and hung over the canyon, which gives a chance to the visitors to enjoy the beauty of the canyon while swimming and having fun.

The hotel rooms start from the -6.60 level. The VIP rooms are situated on the last 2 floors, have large balconies that are hung over the canyon as well.

A large restaurant is included, the entrance of which is on the mark +6.600. The restaurant also has different approaches from the main road for getting food, for the workers. The kitchen block and the block for workers have their own approaches from the road as well, so the visitors and the workers will not encounter.

On the other side of the bridge is the entrance of the multifunctional part of the hotel. The multifunctional center includes cinema halls, game zones, cafes, stores etc. To become the part of the amazing nature of the city, vertical bonds are designed. There are several elevators which give an opportunity to go down to the canyon under the bridge directly from the hotel. Different terraces are designed as well. They give a chance to visitors to get closer to the nature and to the canyon. Tennis courts, football fields, cafes and little parks are located on that terraces making them unique recreation parts for the city.

三等奖 Third Prize

明月桥

The Bright Moon Star Dilute

学校名称 University (College) Name
天津城建大学，中国
Tianjin Chengjian University, China

指导教师 Supervisor(s)
薛 江 Xue Jiang

参赛学生 Participant(s)
张 兵 Zhang Bing
王金福 Wang Jinfu
黄子文 Huang Ziwen
赵 硕 Zhao Shuo
芦春龙 Lu Chunlong
赵全昌 Zhao Quanchang

设计说明 Design Description

本设计以西部山区某高速公路工程为背景。桥位以山谷地形为主，采用中承式拱桥。桥梁全桥长 120m，设计车道为双向 6 车道；主拱券采用钢管混凝土结构，拱券横撑采用不同形状的月牙形状。该桥从外观看主拱券像弯月，拱券之间的横撑月牙结构代表着月亮圆缺变化。故该拱桥取名为明月桥。

This design takes a highway project in the Western Mountain Area as the background. The bridge location is mainly valley topography and the half-through arch bridge is adopted. The whole length of the bridge is 120m and its design lane is two-way 6 lanes; the main arch ring adopts concrete-filled steel tube structure, and the transverse brace of the arch ring adopts different crescent shapes. From the outside, the main arch ring of the bridge looks like a crescent moon, and the crescent structure between the arch rings represents the change of the moon. Based on the above analysis , the arch bridge is named The Bright Moon Star Dilute.

■ 优秀奖 Honorable Mention

鱼跃苇上

Fish Over Reeds

学校名称 University (College) Name
北京建筑大学，中国
Beijing University of Civil Engineering and Architecture, China

指导教师 Supervisor(s)
焦驰宇 Jiao Chiyu
乔 宏 Qiao Hong

参赛学生 Participant(s)
刘 洋 Liu Yang
程冕洲 Cheng Mianzhou
张湘卓 Zhang Xiangzhuo
李杨杰 Li Yangjie
张家辉 Zhang Jiahui

设计说明 Design Description

此次比赛的“鱼跃苇上”桥梁设计项目，选取雄安新区启动区为背景进行设计。在建筑设计方面，此次设计的桥梁彰显了雄安新区的科技与文化，“鱼”跃“苇”上的设计理念呈现出来一幅美妙的白洋淀画卷，也彰显了对雄安新区的美好祝福。在结构体系方面本桥结构造型简洁、轻巧，采用独具特色的结构样式，且结构受力合理，是一个合格的设计方案。

The “Fish Over Reeds” bridge was designed with the Xiong'an New Area as the background. The bridge design highlights the technology and culture of the Xiong'an New Area, and the design concept of “fish jumping over reeds” presents a wonderful painting of the Baiyangdian, as well as a blessing to the Xiong'an New Area. In terms of structural system, the bridge is simple and lightweight, with a unique structural style and reasonable structural forces, making it a qualified design solution.

■ 优秀奖 Honorable Mention

叠 桥

Die Qiao

学校名称 University (College) Name
重庆交通大学，中国
Chongqing Jiaotong University, China

指导教师 Supervisor(s)
董莉莉 Dong Lili　　周志祥 Zhou Zhixiang

参赛学生 Participant(s)
夏宇晨 Xia Yuchen　　冯 麟 Feng lin
施芋宇 Shi Yuyu　　钟星星 Zhong Xingxing
张箴虹 Zhang Zhenhong

设计说明 Design Description

本桥取名为叠桥，取自中国成语“遇水叠桥”。意思是遇水阻拦，就架桥通过。形容不怕阻力，奋勇前进。

桥梁选址在意大利北部阿尔卑斯山下如诗如画的科莫湖北部。科莫湖气候多变，风浪较大，而湖边的景色宜人，景点有较多游客，然而无法有序组织。

该设计分为两层，呼应了名字“叠”，组织由三个方向交汇而来的人流。在风景秀丽的科莫湖上形成一道景观线。

The bridge is named Die Qiao, which is taken from the Chinese idiom “build bridges across the rivers”. This means crossing the bridge when it is blocked by water, decribing the spirity of being unafraid of resistance and move forward bravely.

The bridge is located in the northern part of the picturesque Lake Como in the Alps of northern Italy. Lake Como has a changeable climate, with high winds and waves, but the lakeside scenery is pleasant, so there are many tourists, but it cannot be organized in an orderly manner.

The design is divided into two layers, echoing the name “stack” which expresses “die” by Chinese phonetic alphabet, the people flows from three directions meet on the bridge. A landscape line is formed on the scenic Lake Como.

优秀奖 Honorable Mention

垂虹桥

Hung Hom Bridge

学校名称 University (College) Name
北京建筑大学，中国
Beijing University of Civil Engineering and Architecture, China

指导教师 Supervisor(s)
焦驰宇 Jiao Chiyu　　乔 宏 Qiao Hong

参赛学生 Participant(s)
刘 洋 Liu Yang　　程冕洲 Cheng Mianzhou
张湘卓 Zhang Xiangzhuo　　李杨杰 Li Yangjie
张家辉 Zhang Jiahui

设计说明 Design Description

垂虹桥的寓意取自于彩虹世界。桥两端的圆弧使桥看上去像天边的彩虹，而其倒影就在水面。“雾开”在满语中是鲫鱼的意思，弯起的拱有如鱼之身龙之脊，恰似鱼跃龙门。“垂虹”意为垂落尘世的彩虹。引用了李白的诗句“两水夹明镜，双桥落彩虹”。在选色上采用了中国红和莹白的搭配，富有中国画意境。在结构分析方面，我们采用 midas Civil 软件进行建模分析，并通过 midas Civil Designer 软件进行后处理验算，项目安全可靠，能够投入使用 50 年。在满足规范要求的同时又兼观赏性，充满人文气息，真正做到了人文、环境和社会友好。

The Hung Hom Bridge takes the meaning of the rainbow of the world. The arch on both sides of the bridge is like a rainbow in the sky, and its shape is reflected in the water. The bridge crosses the fog to open the river, “Wukai” in the Manchu language is crucian carp meaning, we associate with the fish jumps into the dragon gate, the curved arch has such as the fish body and dragon ridge. In color selection we also used the Chinese classical red and white with rich in artistic conception of Chinese painting. On the structure calculation, we use the midas civil to carry on the modeling analysis, use the software midas Civil Designer to carry on the post-processing checking calculation, satisfies the specification request simultaneously. And also ornamental, full of humanistic atmosphere, truly achieved the humanities, environment and social friendliness.

■ 优秀奖 Honorable Mention

弦歌桥

Xiange Bridge

学校名称 University (College) Name
山东建筑大学，中国
Shandong Jianzhu University, China

指导教师 Supervisor(s)
司 炜 Si Wei　　陈 静 Chen Jing

参赛学生 Participant(s)
Ilunga Stephane Lavery　　刘 旭 Liu Xu

设计说明 Design Description

桥梁设计中，通过方案比选，确定出适合复杂地质条件下的双塔双索面斜拉桥设计方案，并随之展开设计、分析、计算、绘图、施工规划、效果渲染的全过程，力求内容全面、可靠、新颖。本设计方案还注重在安全、创新、可施工、可管养多方面有所突破，例如在大体积混凝土温控技术、软土地基基础选择、合拢口劲性骨架施工管控等关键环节上有所研究及优化设置，体现出较强的社会应用价值和实践特色。

In the bridge design, through the comparison and selection of schemes, the design scheme of the double-tower double-cable surface and cable-stayed bridge which is suitable for complex geological conditions is determined, and then the whole process of design, analysis, calculation, drawing, construction planning, and effect rendering is carried out to make the design content comprehensive, reliable and novel. Pay more attention to breakthroughs in safety, innovation, workability, and manageability, such as research and optimize settings in key links such as mass concrete temperature control technology, soft soil foundation selection, and construction control of closed stiffness framework have been finished to reflect strong social application value and practical characteristics.

■ 优秀奖　Honorable Mention

玉带桥

Jade Belt Bridge

学校名称　University (College) Name
山东建筑大学，中国
Shandong Jianzhu University, China

指导教师　Supervisor(s)
张可心　Zhang Kexin　　姚　越　Yao Yue

参赛学生　Participant(s)
胡佳赢　Hu Jiaying　　谭福利　Tan Fuli
安　芳　An Fang　　孙欣欣　Sun Xinxin

设计说明 Design Description

本设计为人行景观桥，桥位位于西安世博园内。以“丝绸”为造型灵感，有“连结历史，烛照未来”之意。取名“玉带桥”，远观如同在空中飘动的玉带。桥面由三个相连的曲面组合而成，三条路径在中间互通，行人可以从其中一条走到另外任意两条路上。桥两侧护栏采用钢化夹胶玻璃材质，安全且景观性较好。

玉带桥采用标准跨径为150m=55m+40m+55m三跨、等截面连续钢箱梁桥，桥面宽度为3.7m+2.6m+3.7m=10m。

This work is a pedestrian landscape bridge design, which is located at Xi'an Expo Park. The design is inspired by “silk cloth”, which means “connecting the past, lighting up the future”. It is named “Jade Belt Bridge”. It looks like a jade belt floating in the air from a distance. The bridge deck is composed of three interlinked curved surfaces. Three roads interconnect in the middle, and pedestrians can walk from one to the other two roads. The guardrails on both sides of the bridge are made of toughened laminated glass which is safe and provides a good view of landscape.

The Jade Belt Bridge is built into a three-span and equal-section continuous steel box girder bridge whose standard span is 150m=55m+40m+55m and the bridge deck width is 3.7m+2.6m+3.7m=10m.

■ 其他作品 Other Works

合谊之桥

Bridge of Friendship

学校名称 University (College) Name
安徽建筑大学，中国
Anhui Jianzhu University, China

指导教师 Supervisor(s)
周 宇 Zhou Yu
陈 东 Chen Dong

参赛学生 Participant(s)
孙超越 Sun Chaoyue
王 浩 Wang Hao
奚圣宇 Xi Shengyu
徐焱明 Xu Yanming
张贺强 Zhang Heqiang

设计说明 Design Description

此景观桥长 200m，宽 6m，其中桁架钢桥 100m，两侧接混凝土桥各 50m。钢桥部分采用圆钢管空腹桁架中承受力结构，顶部的两端斜拉八根拉索，连接到混凝土桥梁处，拉索平衡上弦压杆，形成自平衡体系；下部采用贝雷桁架片， 横桥面为大面板钢板满铺，两侧安装木制护栏。混凝土桥采用鱼腹箱型截面，采用自愈合混凝土铺设。桥梁采用钢桥与廊桥结合，混凝土桥与钢架桥结合的方式，体现了文化交融之美，各国在文化的交流中共同发展，实现共赢。

The overall landscape bridge is 200m long and 6m wide, comprising 100m of truss steel bridge with 50m of concrete bridge on both sides. The steel bridge adopts the circular steel pipe vierendeel truss middle bearing structure, with eight oblique cables from the top connecting the concrete bridge. The cable balances the upper chord compression bar, thus forming a self-balance system. The ground of the bridge adopts the Bailey truss pieces, and the transverse bridge deck is covered with the large deck steel plates, at which the wooden guardrail is installed on both sides. The concrete bridge representes fish belly box section and self-healing concrete is applied to the construction. It is the combination of steel bridge and corridor bridge, concrete bridge and steel frame bridge, which shows great glamours of cultural communication and a win-win cooperation exists in mutual communication for all the people around the world.

■ 其他作品 Other Works

尧山服务区匝道

Ramp for Yaoshan Expressway Service Area

学校名称 University (College) Name
河南城建学院，中国
Henan University of Urban Construction, China

指导教师 Supervisor(s)
屈讼昭 Qu Songzhao　　王 仪 Wang Yi
董召锋 Dong Zhaofeng

参赛学生 Participant(s)
孙 涛 Sun Tao

设计说明 Design Description

尧山服务区 F 匝道桥上跨预留 A、C 匝道，桥梁中心桩号为 FK0+755.000，桥梁起点桩号为 FK0+626.800，桥梁终点桩号为 FK0+883.200，桥梁全长为 256.4m。桥梁跨径组合为：第一联（25+30+25+25）m，第二联（30+35+30）m，第三联（25+25）m，等截面预应力混凝土现浇连续箱梁，桥墩墩身采用柱式桥墩（圆形截面扩头）；0 号、9 号桥台采用肋式台；全桥基桩均采用钻孔灌注桩。本桥平面位于 A=350m 的缓和线段、R=800m 的左偏圆曲线和 A=350m 的缓和线段内，墩台径向布置。纵面依次位于 i_1=3.45% 的直坡段，前坡 i_1=3.45%、后坡 i_2=−3.50% 的凸曲线、i_3=−3.50% 的曲线段（R=2600m、T=90.35m、E=1.579m），变坡点桩号、高程为 FK0+745.000 高程为 55.118m。

Ramp A and Ramp C is reserved across the Ramp F bridge in Yaoshan Service Area. The 256.4m-long bridge has a central chainage FK0+755.000, a start chainage FK0+626.800, and an end chainage FK0+883.200. The combination of spans of the bridge is as follows: the first span (25+30+25+25)m, the second span (30+35+30)m, and the third span (25+25)m, with the same section prestressed concrete cast-in-place continuous box girder, and the column type pier (circular section extension head) shall be used for pier body; Rib abutment shall be used for 0# and 9# abutments. Cast-in-place bored pile shall be used for the foundation piles in the whole bridge. The bridge plane is located in the moderate line segment of A=350m, the left circular curve of R=800m and the moderate line segment of A=350m. The pier is arranged in the radial direction. The vertical plane is successively located in the straight slope section of i_1=3.45%, the convex curve with front slope of i_1=3.45% and rear slope of i_2=−3.50%, and the curved section of i_3=−3.50% (R=2600m, T=90.35m and E=1.579m). The chainage and elevation of grade change point are respectively FK0+745.000 and 55.118m.

■ 其他作品 Other Works

佑安大桥

You An Bridge

学校名称 University (College) Name
吉林建筑大学，中国
Jilin Jianzhu University, China

指导教师 Supervisor(s)
闫 铂 Yan Bo　　温永钦 Wen Yongqin

参赛学生 Participant(s)
张 挺 Zhang Ting

设计说明 Design Description

佑安大桥，意为保佑平安，桥梁的建立使人造美与自然美交相辉映，也增加了城市的立体感。大桥全长 203m，是一座双向四车道的钢箱桥。桥梁的建设因地制宜，由于大桥需要跨过一条高速公路，为了避免施工对公路交通造成影响，对桥梁跨中部分采用顶推法施工。此技术具有施工速度快、安全性能高、材料消耗小等优点，大大提高了桥梁的性价比。佑安大桥的建成，将会缩短市民的通行时间，也将会取得良好的社会价值和经济效益，全面打造城乡一体化新格局。

The You An Bridge, means stable and peaceful, aims to protect this pure land and benefit the local people. The combination of artificial beauty and natural beauty created by the bridge enhances the three-dimensional sense of the city. With a total length of 203 meters, the bridge is a two-way four lanes' steel box bridge. The construction of the bridge should be adapted to the local conditions. As the bridge needs to cross a highway, in order to avoid the impact of the construction on the highway traffic, the jacking method is used for the construction of the middle part of the bridge span. This technology has the advantages of fast construction speed, high safety performance and low material consumption, which greatly improves the performance to price ratio of the bridge. The completion of You An bridge will not only shorten the passage time of citizens, but also achieve good social value and economic benefits, and build a new pattern of urban-rural integration in an all-round way.

■ 其他作品 Other Works

丝梦大桥

Silk Dream Bridge

学校名称 University (College) Name
吉林建筑科技学院，中国
Jilin University of Architecture and Technology, China

指导教师 Supervisor(s)
李　强 Li Qiang

参赛学生 Participant(s)
曹明昊 Cao Minghao
郭　渭 Guo Wei
侯占辉 Hou Zhanhui
纪庆玲 Ji Qingling
王旭龙 Wang Xulong

设计说明 Design Description

桥梁名称为丝梦大桥，源于“一带一路”倡议。丝绸之路经济带和 21 世纪海上丝绸之路，为沿线各国搭建起中国梦与世界梦息息相通的桥梁，谱写新的追梦华章。是顺应世界多极化、经济全球化、文化多样化、社会信息化的潮流。秉持开放的区域合作精神，致力于维护全球自由贸易体系和开放型世界经济。丝梦大桥不仅仅是大桥作用的象征，而且还是新型国际关系的象征。

以丝梦大桥为标志性建筑，弘扬丝路精神。作为人文社会的交往平台，多民族、多种族、多宗教、多文化在此交汇融合，形成团结互信、平等互利、包容互鉴、合作共赢的新局面。不同种族、不同信仰、不同文化背景的国家可以共享和平，共同发展。

The bridge is named Silk Dream Bridge, which is derived from the Belt and Road Initiative. The Silk Road Economic Belt and the 21st Century Maritime Silk Road have built a bridge between the Chinese dream and the world's dream, writing a new chapter in pursuit of dreams. It is to conform to the trend of world multi-polarization, economic globalization, cultural diversification, and social informatization. The Silk Dream Bridge is not only a symbol, but also a representation of the new type of international relations.

The Silk Dream Bridge will stand as a landmark building to promote the spirit of the Silk Road. As the communication platform of the humanistic society, people of different ethnic groups, races, religions, and

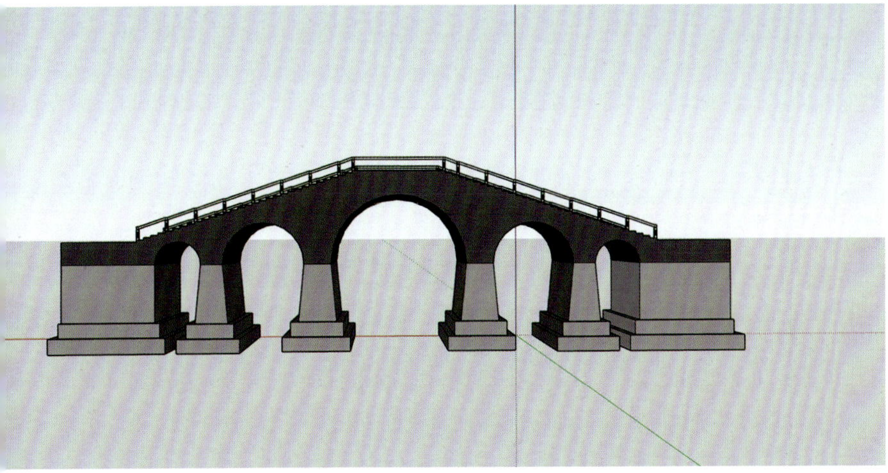

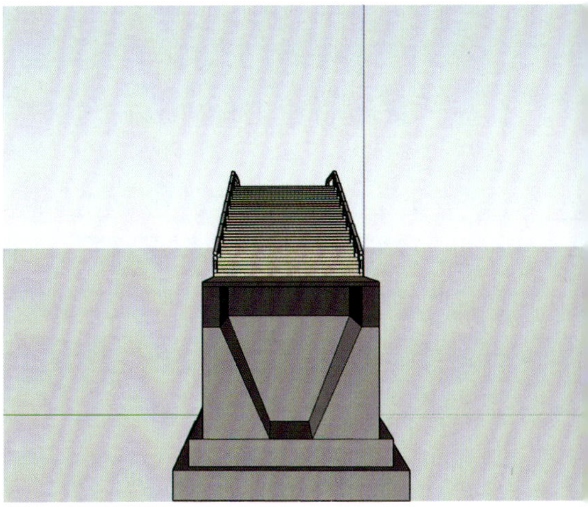

cultures are gathered here. The Silk Road spirit has formed that based on unity and mutual trust, equality and mutual benefit, tolerance and mutual learning, and win-win cooperation, countries of different races, beliefs and cultural backgrounds can share peace and develop together.

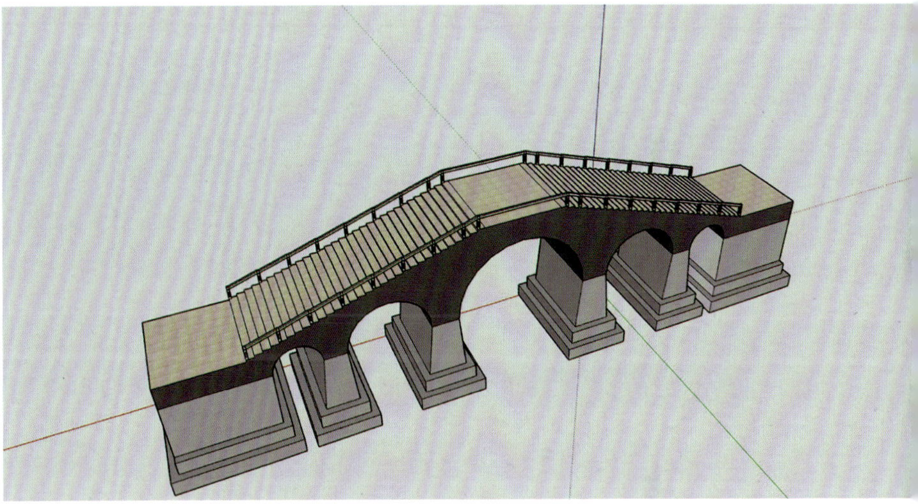

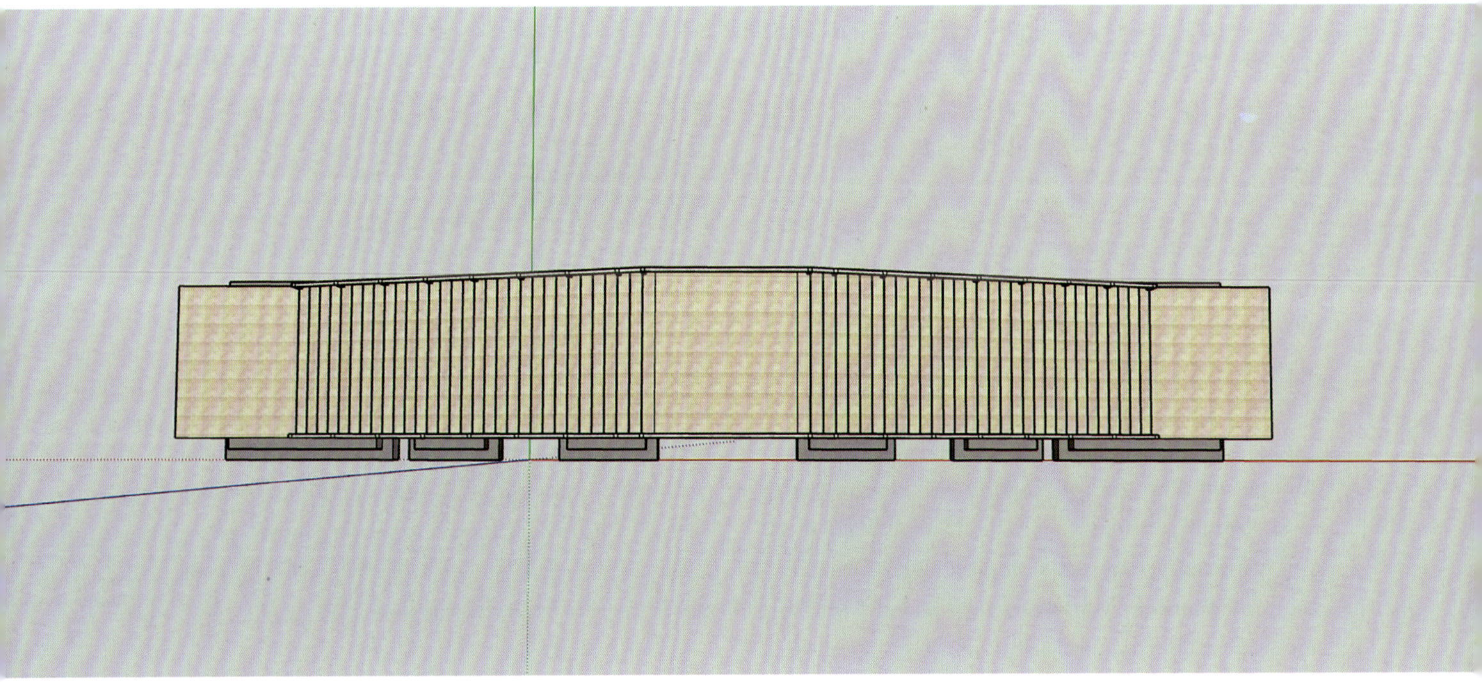

■ 其他作品 Other Works

“海上丝绸”之桥

“Silk on the Sea” Bridge

学校名称 University (College) Name
吉林建筑科技学院，中国
Jilin University of Architecture and Technology, China

指导教帅 Supervisor(s)
李 强 Li Qiang

参赛学生 Participant(s)
曹明昊 Cao Minghao
郭 渭 Guo Wei
侯占辉 Hou Zhanhui
纪庆玲 Ji Qingling
王旭龙 Wang Xulong

设计说明 Design Description

随着时代发展，丝绸之路成为古代中国与西方所有政治、经济、文化往来通道的统称。包含“陆上丝绸之路”“海上丝绸之路”及“草原丝绸之路”。“一带一路”倡议致力于亚欧非大陆及附近海洋的互联互通，建立和加强沿线各国互联互通伙伴关系，构建全方位、多层次、复合型的互联互通网络，实现沿线各国多元、自主、平衡、可持续的发展。“一带一路”倡议的互联互通项目将推动沿线各国发展战略的对接与耦合，发掘区域内市场的潜力，促进投资和消费，创造需求和就业，增进沿线各国人民的人文交流与文明互鉴，让各国人民相逢相知、互信互敬，共享和谐、安宁、富裕的生活。建设“海上丝绸”之桥架起全球人民之间的桥梁。

With the development of the times, the Silk Road has become the collective symbol of all the political, economic and cultural exchanges between China and the West. “Land Silk Road” “Maritime Silk Road” and “Grassland Silk Road” are included. The Belt and Road Initiative aims to promote the connectivity of Asian, European and African continents and their adjacent seas, establish and strengthen partnerships among the countries along the Belt and Road, set up all-dimensional, multi-tiered and composite connectivity networks, and realize diversified, independent, balanced and sustainable development in these countries. The connectivity projects of the Initiative will help align and coordinate the development strategies of the countries along the Belt and Road, tap market potential in this region, promote investment and consumption, create demands and job opportunities, enhance people-to-people and cultural exchanges, and mutual learning among the peoples of the relevant countries, and enable them to understand, trust and respect each other and live in harmony, peace and prosperity. Building the “Silk on the Sea” Bridge is to connect people around the world.

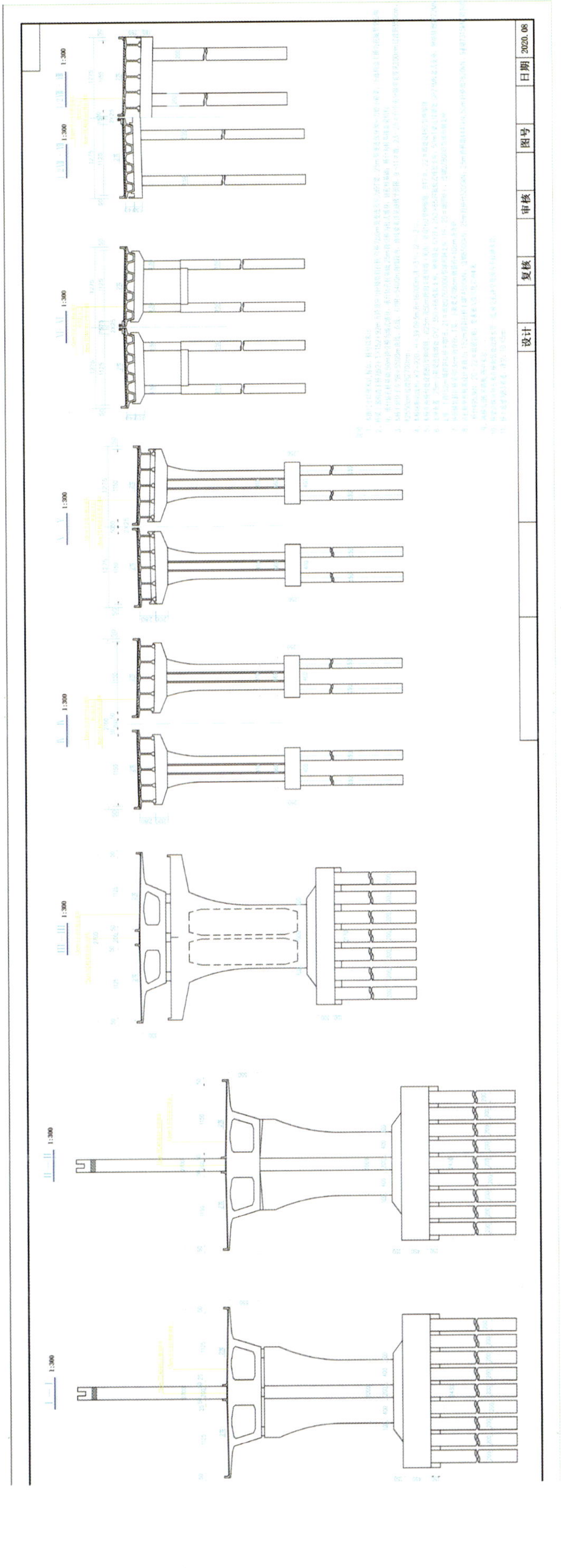
日期
2020.08
图号
审核
复核
设计

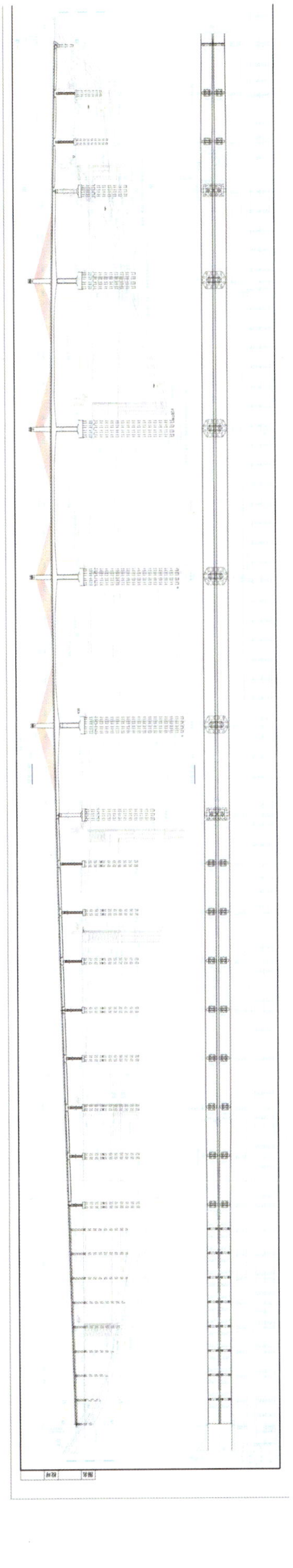

■ 其他作品 Other Works

Myasnikyan 大道休闲区交通桥

Bridge with urban development concept of recreation areas along Myasnikyan avenue

学校名称 University (College) Name
莫斯科国立建筑大学，俄罗斯
Moscow State University of Civil Engineering, Russia

指导教师 Supervisor(s)
Kolotovichev Yuri　Smirnov Vladimir

参赛学生 Participant(s)
Svetlana Babiy　Vitaly Zykov
Zaurbek Gadaborshev　Valeria Maslova

设计说明 Design Description

Myasnikyan 大道的诸多休闲区之间并没有相互连接。此外，当地有大量未开发的土地和自然景观俨然已成为特色，选择修建桥梁可以避免破坏这些景观。

该地区的主要问题之一是当前的道路网络已无法满足交通需求，出现这种情况，一是因为交通方式过于单一，二是因为埃里温的道路并不通达。根据埃里温的总体发展规划，未来的交通路线在穿过 Davit Anhaght 街和 Rubinyants 街后将继续向 Nork 延伸，连接 Nor Nork。本方案设计的交通桥便是埃里温的 Kentron 行政区的旁路之一，高 10m，主要分担 Myasnikyan 大道的交通负荷。

另外，索道也是备选的交通方案，可以用于连接如亚美尼亚国家歌剧院和芭蕾舞剧院和植物园等地点。

计划在 Davit Anhaght 街和 Ulnetsi-Rubinyants 街的交叉口建造多功能枢纽和交通桥，交通桥将连接两个峡谷景点，桥下可以修建电影院、露天剧场、公园等设施。

The importance of the design is in the fact that a number of recreation areas along Myasnikyan Avenue do not have a common connection. Most of the area is unused uncultivated areas, as well as there are various Architectural style and significant structures that distort the appearance of the natural landscape.

One of the main problems of the area is the road network, which does not meet the requirements of the current traffic. This problem arose mainly due to the lack of alternative means of transport for fast movement, as well as the lack of a road bypassing Yerevan. According to the general plan of Yerevan, after crossing Davit Anhaght Street with Rubinyants Street, it should continue to Nork and branch off and also join Nor Nork. The design proposes to build a road bypassing the Kentron administrative district of Yerevan and unloading the traffic on Myasnikyan Avenue at a height of 10m from Myasnikyan Avenue.

As an alternative means of transportation, it is proposed to build a ropeway, due to which it will be possible to reach the Botanical Garden from the building of the National Opera and Ballet Theater of Armenia without the help of a car.

It is also planned to build a new multifunctional center and bridge at the intersection of Davit Anhaght and Ulnetsi-Rubinyants streets. The new bridge will connect the two parts of the gorge. The bridge will have cinemas, amphitheaters and parks under it.

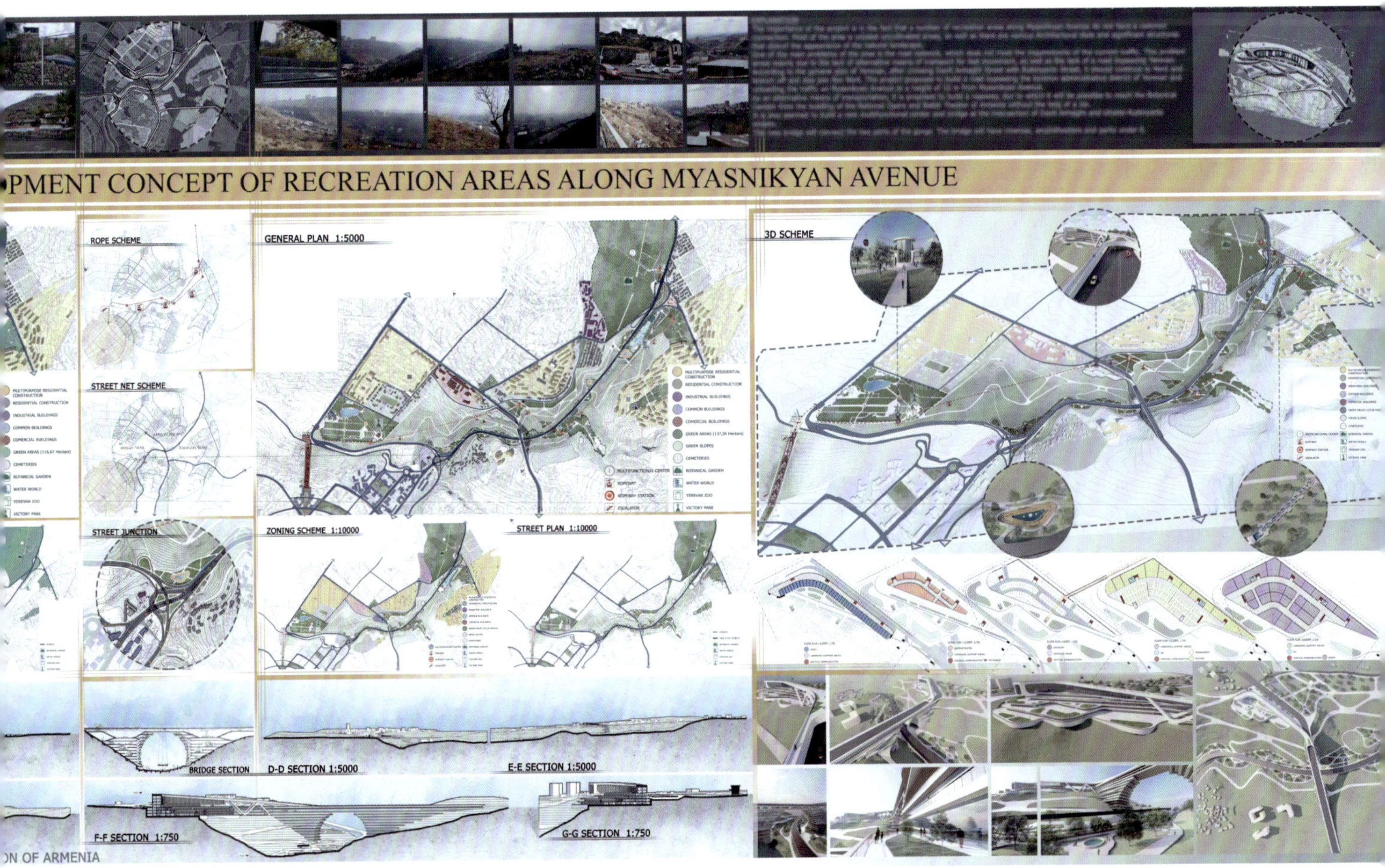

其他作品 Other Works

国风桥

Guofeng Bridge

学校名称 University (College) Name
山东建筑大学，中国
Shandong Jianzhu University, China

指导教师 Supervisor(s)
司 炜 Si Wei　　陈 静 Chen Jing

参赛学生 Participant(s)
Bipfubusa Miamoneka Steffy

设计说明 Design Description

设计中秉持兼顾“安全、适用、耐久、创新、美观、节约、可施工、可管养”等理念，依托复杂地质条件，完整实现一大跨悬索桥的设计、分析、计算、绘图、施工规划、效果渲染。全过程力求内容全面、可靠、新颖。本设计方案更在美学设计、耐腐蚀新工艺、数字化智能管理与维护等多方面有所研究及优化设计，体现出较强的社会应用价值和实践特色。

The design adheres to the concepts of “safety, applicability, durability, innovation, beauty, economy, workability, and manageability”, and relying on complex geological conditions, the design, analysis, calculation, drawing, construction planning, and construction of a large-span suspension bridge are fully realized. The whole process of effect rendering strives to be comprehensive, reliable and novel in content. This design plan has researched and optimized design in many aspects of aesthetic design, new corrosion-resistant technology, digital intelligent management and maintenance, which reflects strong social application value and practical characteristics.

■ 其他作品 Other Works

涌泉桥

Yong Quan Bridge

学校名称 University (College) Name
山东建筑大学，中国
Shandong Jianzhu University, China

指导教师 Supervisor(s)
曹三鹏 Cao Sanpeng

参赛学生 Participant(s)
孟祥伦 Meng Xianglun
聂配程 Nie Peicheng
肖鸿儒 Xiao Hongru
周浩然 Zhou Haoran

设计说明 Design Description

设计方案：斜拉桥

本设计以斜拉桥设计基本理论和静－动力分析为理论基础，以成功修建的斜拉桥（如黄河大桥）为例子，由于斜拉桥为超静定结构，计算比较烦琐，我们决定采用专业设计软件 MAIDAS、REVIT 广厦 CAD 等软件进行结构内力组合以及位移比地震周期的检核，以及判定设计的合理性。

设计洪水频率：极低。

设计行车速度：100km/h。

桥梁所处地区地形：平原地带，地形平缓；地质结构稳定。

气候气象：温带季风气候。夏冬季常有强对流天气。

Design scheme: Cable-stayed bridge

This design is based on the basic theory of cable-stayed bridge design and static and dynamic analysis. Taking the successfully built cable-stayed bridge (such as the Yellow River Bridge) as an example, because the cable-stayed bridge is an ultra-statically determinate structure, the calculation is relatively cumbersome, we decided to use the professional design software MAIDAS, REVIT Guangsha CAD and other softwares to check the internal force combination of the structure and the seismic cycle of the displacement ratio, and determine the rationality of the design.

Design flood frequency:Very low.

Designed driving speed: 100km/h.

Terrain of the area where the bridge is located: plain area, gentle terrain; stable geological structure.

Climate and Meteorology: Temperate monsoon climate. Strong convection weather in summer and winter.